D1251989

THE WIRELESS DATA HANDBOOK

James F. DeRose

Editor and Publisher *Paul Mandelstein*
Project Management, Electronic Composition *Professional Book Center*

Published by Quantum Publishing, Inc.
P.O. Box 310
Mendocino, California 95640

Library of Congress Cataloging-in-Publication Data
is available for this title

ISBN 0-930633-19-9
Printed in the United States of America

CONTENTS

PART

I

GETTING STARTED

CHAPTER 1

A SHORT HISTORY OF DATA RADIO

1.1 IN THE BEGINNING...

In 1899, four years after Marconi's first wireless telegraph, the British Navy converted to data radio.[1] The czar's navy quickly followed. By 1905 the Japanese had mastered the key techniques and began to intercept messages from the Russian Vladivostok fleet cruising secretly south of Tokyo Bay. The victorious (for Japan) Battle of Tsushima followed.

Driven by continued military demands, wireless data technology leaped forward. By 1907 it was in common use on ships; in 1914 the hapless Russians lost the Battle of Tannenburg because of German intercepts of their land-based data radio communications;[2] in 1917 the British successfully employed radio telegraph in tanks at the Battle of Cambrai;[3] by 1918 these same units were adapted for aircraft.

In World War II both the United States and Germany communicated with and controlled their widely scattered submarine fleets via data radio. During this period H. C. A. Van Duuren[4] devised the technique still known as ARQ—Automatic Repeat reQuest—one of those disarmingly simple ideas that seems trivial in retrospect. The idea was to ensure that a block of characters had been successfully transmitted through the use of error *detection*. A detected error was followed by a signal from the receiver asking the transmitter to repeat the block.

In 1956 the SAGE Air Defense System began testing digitized radar information sent by data radio from airborne early warning aircraft and "Texas Towers." The more complex manually initiated messages—angle of elevation, slant range—were separated from the continuously repeating X and Y coordinates. A header identifying the radar address was added to this key data. These manageable segments were clear forerunners of what would later be called "packets." These packets traveled over dedicated circuits as in today's data-over-cellular.

In 1957 data radio modems reached speeds of 2000 bits per second (bps), competitive with landlines; by 1967 the General Dynamics ANDEFT/SC-320 modem achieved 4800 bps.[5] The stage was set for commercial exploitation of this knowledge base.

1.2 PRIVATE NETWORKS LEAD THE WAY

In 1969 IBM began to develop a mobile data radio system for police departments. Fueled by block grants from the Law Enforcement Assistance Administration (LEAA), other vendors, including Kustom, Motorola, Sylvania/LTV, and Xerox, offered alternatives. IBM's 2976 Mobile Terminal System was announced on May 12, 1972 (see Appendix B). A polled system, it nonetheless achieved good throughput with a combination of high (~5400 bps) airlink transmission speeds, ½ rate forward error correction, and ARQ.

The IBM system was a failure and was withdrawn in 1974. Competition fared little better. The failure causes were many:

1. The termination of LEAA funding in 1973
2. The physical inadequacy of the devices: big, heavy, hot, noisy, and unreliable
3. A crushing lack of software support: no dispatch applications, etc.
4. Unreadiness of the customers: no databases or applications in place, etc.

But the dreamers persisted. In 1970 the successful University of Hawaii's ALOHA system had established fundamental inbound contention techniques.[6] In 1975 Kleinrock and Tobagi[7] codified carrier sense multiple access (CSMA) techniques, permitting greatly improved inbound performance. MDI, founded in 1978 to provide a data radio system for the Vancouver (B.C.) police, began to work with Federal Express in 1979. The following year the first 12 commercial terminals employing CSMA were delivered.

Meanwhile, IBM's Service Research organization had been privately piloting briefcase-size "portable" radio terminals, developing a business case for the applications that would yield economic payoff. In November 1981 a contract was signed with Motorola for the Digital Communication System (DCS). Nationwide rollout began in April 1984 and was essentially complete two years later with the installation of more than 1,000 base stations.

DCS was a pedestrian-oriented system that broke much new technical ground. It used a single frequency on adjacent base stations, with deliberate overlapping coverage patterns, to achieve better in-building penetration and reduce contention impact. And the end-user device was handheld, incorporating radio modems and internal dual-diversity antennas for improved reception at walk speeds.

1.3 THE RISE OF PUBLIC NETWORKS

Of equal business importance, with the signing of the DCS contract IBM and Motorola agreed to work together on a shared network approach. The initial opportunity estimates were enormous, the first era of "low-hanging fruit."

But during the period 1983–1985 there were serious business disagreements between the two equally proud companies. IBM better understood the application development gates that would hinder rapid rollout of this technology and had placed experimental customers on DCS via the IBM Information Network. Motorola had a sounder grasp of the infrastructure changes necessary to provide a high availability system and had begun development of its own network. The proposal to build a public network resting on the shoulders of DCS was rejected by the decision-making elements of both IBM and Motorola for complex internal business reasons.

After the collapse of the joint venture negotiations, Motorola unveiled its own public packet network, the Digital Radio Network (DRN) based on modified DCS technology. It began in Chicago in 1986 (Ericsson began Mobitex in Sweden the same year). As IBM expected, making a market was an extraordinarily difficult task. It was four years before DRN-Chicago had ~135 external users;[8] Los Angeles, rolling out second, took 20 months to achieve about the same number; New York City, 18 months behind Los Angeles, reached the 135 milestone after one year. Clearly a learning curve existed, but the absolute pace was exceedingly slow.[9]

Five years after the negotiations failed, the plan was refurbished (Motorola's DRN was incorporated), agreed by both companies, and the Advanced Radio Data Information System (ARDIS) was announced in January 1990. The system has been in continuous evolution ever since: new devices; a new, higher bit-rate protocol; additional frequencies; roaming capability; and extraordinary redundancy added to achieve high availability. The customer base has grown, slowly, but increasing it remains a continuous struggle.

In October 1990 RAM announced its public data service based upon Ericsson's newest Mobitex design. Initially plagued by a lack of nearly everything—adequate base stations/coverage, no handheld modems—RAM's lack of success at achieving an adequate subscriber base has been extraordinarily painful. To date RAM has achieved only 1–2 subscribers per base station, but many promising business relationships. Nationwide coverage was achieved in June 1993; it is clear this year will be pivotal for the business partners.

Meanwhile, noticeable progress was made with data-over-cellular. Users began to dial connections on this circuit-switched facility much as they used ordinary landlines—greatly helped by far better modem capability. Portable facsimile alone created thousands of casual public data radio users.

Further, selected high-volume customers made unusual connections to the cellular mobile telephone switching offices (MTSOs), bypassing the public switched network landline portion for direct connection to its own leased facilities. By February 1993 United Parcel Service (UPS) had approximately 50,000[10] operational units with this reduced tariff arrangement, significantly more subscribers than DRN/ARDIS had after seven years of effort.

Multiple public network alternatives are being readied for 1994, most notably cellular digital packet data (CDPD) implementations by multiple carriers, as well as other truly digital (requiring no modem) offerings. The existence of this enormous excess capacity is certain to drive prices down; indeed, price discounting by extant service providers has already begun. Perhaps—at last—the long awaited, widespread adoption of wireless data is about to be triggered.

REFERENCES

[1] *The Telecommunications Industry*, Gerald W. Brock, p. 162.

[2] *Great Battles of World War I*, Anthony Livesy, p. 31.

[3] Ibid., p. 128.

[4] "Computer Communications: Milestones," Paul E. Green, Jr., *IEEE Communications Magazine*, May 1984, p. 55.

[5] "Error Distribution and Diversity Performance of a Frequency-Differential PSK HF Modem," Gene C. Porter, *IEEE Transactions on Communications Technology*, August 1968, pp. 567–575.

[6] "The ALOHA System—Another Alternative for Computer Communications," N. Abramson, *AFIPS Conference Proceedings*, vol. 37 (Montvale, NJ: AFIPS Press, 1970, pp. 281–185).

[7] "Packet Switching in Radio Channels, Part I: Carrier Sense Multiple Access Modes and Their Throughput-Delay Characteristics," L. Kleinrock and F.A. Tobagi, *IEEE Transactions on Communications Technology*, December 1975, pp. 1400–1416.

[8] From FCC loading records.

[9] It is worth noting that Mobitex Sweden has fared worse. After eight years the total subscriber count is ~8,000. Source: Ovum, Ltd., as reported in *Communications Week International*, 6-28-93.

[10] Paul Heller, UPS Division Manager for Mobile Networks, *Mobile Data Report*, 2-15-93.

CHAPTER

2

DATA NETWORK TYPES

2.1 A ROUGH SORT OF TWO-WAY SYSTEMS

In North America terrestrial, two-way wireless data networks can initially be sorted into private versus public systems. Well-known examples of private systems include Federal Express and most public safety—police and fire department—organizations. In private systems the infrastructure is owned, and the spectrum is licensed, by the user. Public systems are "for hire." The user does not own the infrastructure and has no special rights to the spectrum that carries the traffic. Public systems can be further subdivided into packet switched (e.g., ARDIS, RAM) and circuit switched (e.g., data-over-cellular).

2.2 PRIVATE SYSTEMS

Private systems are usually marked by tight geographic constraints, such as a metropolitan area, a railroad yard, or an airport car rental facility. Even Federal Express is organized on a city basis; package delivery vehicles do not drive from New York City to Boston. Typically, private systems have protected frequencies allocated by the Federal Communications Commission (FCC) for particular industry areas. Further, coverage requirements are usually "street level": relatively few base stations aimed at vehicular targets. Thus, private systems tend to be the provenance of public safety, utility, and taxi/limo dispatch applications.

Very large private systems are vanishing. The constrained availability of spectrum in major metropolitan areas virtually precludes the creation of the vast new 1,000-user-per-city systems like those IBM initiated in 1981. Indeed, IBM's private system has now been absorbed in ARDIS. After 18 months of testing, at the close of 1989 GE Consumer Services selected Kustom Electronics to build its nationwide mobile data network.[1] A year later Kustom began the layoffs[2] that presaged its precipitous decline. In May 1991, with 60 zones operating, GE announced that it would integrate "its existing dispatching operations with RAM Mobile Data . . . to expand its coverage areas and relieve spectrum congestion in some of its cities."[3] GE would "also allow other corporations to use the GE network."

One early vendor motivation for erecting public data systems was that customers would "grow up" and graduate to the vendor's private infrastructure. There is no known example of such an event. Indeed, small private networks such as PSE&G of New Jersey are testing RAM[4] with the serious intent to move to a public system.

Spectrum considerations aside, the economics favoring public systems are compelling. Consider a large metropolitan area requiring 25 base stations to provide the desired coverage. Assume, further, that the seven-year cost for a private network is more than $4 million based on the following:

One-time costs	Qty	Unit cost	Total cost
concentrator/gateway	1	$100,000	$100,000
base stations	25	20,000	500,000
modems	50	1,000	50,000
software licenses			100,000
Seven-year recurring:			
maintenance		14%/year	635,000
leased lines	26	300/month	655,000
rooftop rentals	25	275/month	577,500
Operators	5	25,000 each	875,000
			3,492,500
20% contingency			698,500
			$4,191,000

Then the private versus public break-even curve for this area can be calculated as shown in Figure 2-1. If the user can obtain public service for $75 per month per subscriber, nearly 700 subscribers must exist in this metro area to permit the private sys-

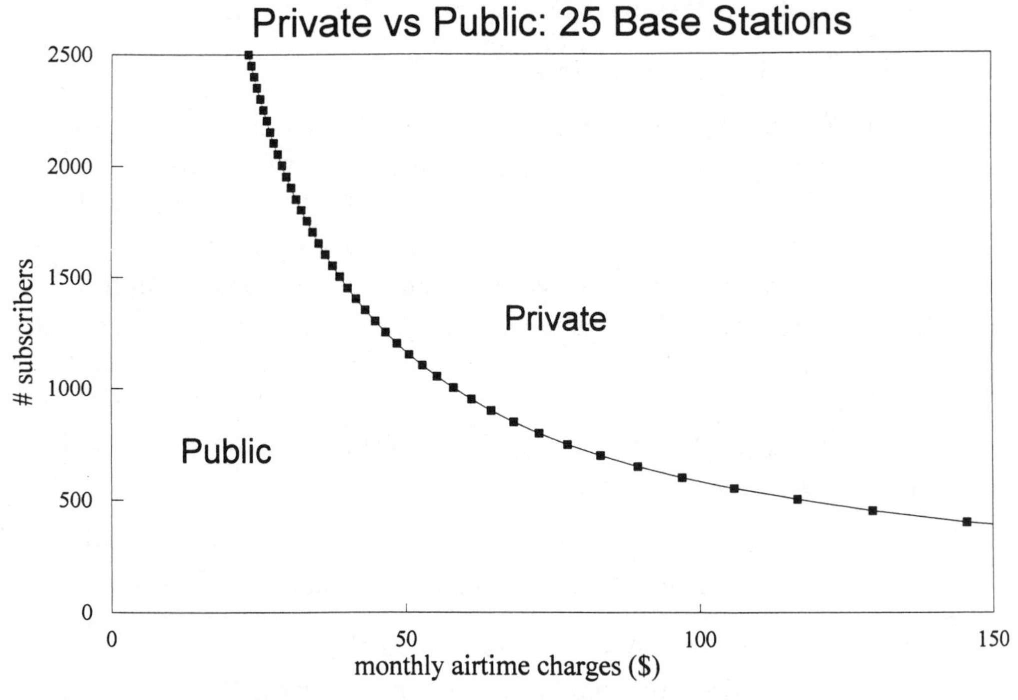

Figure 2-1 *Private versus Public Break-Even Point: 25 Base Stations*

tem to break even in seven years. This would be very unusual. IBM's Field Service force was larger than 750 subscribers in major centers such as, New York City, Chicago, and Los Angeles—an exceptional case. If the user has only 200 subscribers, the private system will cost more than three times as much as going public.

As public networks become ubiquitous, probably a reality by 1995, there will be a tendency for all but public safety users to convert away from their private systems. Even in this bastion the trend toward public systems has begun: the New York City Sheriff's Department switched to ARDIS for parking enforcement,[5] as has Philadelphia.[6] Two of RAM's earliest test customers were the Hoboken (NJ) Fire Department and the Orange County (FL) Sheriff's Department.[7]

2.3 PUBLIC SYSTEMS: DATA-OVER-CELLULAR VERSUS PACKET SWITCHED

2.3.1 Packet Confusion

Nearly 40 years ago the user messages carried by the SAGE system were segmented to provide better system control. During the six-year period from 1958 to 1964 IBM developed the SABRE system for American Airlines, drawing on SAGE knowledge. The message segmentation functions began to be more cleanly defined. The first System/360 teleprocessing system embraced many of these conventions with its basic telecommunications access method (BTAM), but the simple challenge of reconfiguring a line or terminal demonstrated how much work remained.

Gradually structure emerged. Each message segment had a *header* containing basic information that marked at least the

1. Beginning of the message segment
2. Address (often both source and destination)
3. Sequence number

and a *trailer* that held an error-detection code and marked the end of the message segment.

By the time the U.K.'s Donald Davies coined the term *packet* in 1965, a recognizable header/data/trailer structure had been moving across leased, then circuit switched facilities for roughly ten years. "Packets" still flow on circuit switched connections. Their structure often has vendor-proprietary wrinkles to distinguish them from public domain alternatives. But even the flag and the error-detection process used in Microcom's MNP series of protocols are identical to those of the common packet switched networks. A formal packet structure designed for packet switching can be quite at home on circuit switched alternatives. That's what dial backup is all about.

2.3.2 Allocation Alternatives

Two competing approaches to the allocation of transmission bandwidth exist: pre-allocation and dynamic allocation.

Telephone networks are circuit switched systems in which a fixed bandwidth is pre-allocated for the duration of the call—whether it is voice, data, or image. Today's voice cellular systems fall into this category.

Historically, the telegraph dynamically allocated bandwidth one link at a time, never attempting to schedule the whole source-to-destination path. Obviously this service was limited to non-real-time systems. The advent of the computer permitted dynamic allocation techniques to be reexamined for new communication alternatives. Packet switching was the result.

Packet switching was not really an invention, but rather an intelligent reapplication of basic dynamic allocation techniques to data transmission. A packet switched network only allocates bandwidth when a block of data is ready to be sent; sophisticated networks only assign enough bandwidth for one block to travel over one network link at a time.

Depending upon message length, packet switching can be many times more efficient than circuit switching in reducing bandwidth wastage. But packet switching carries relatively high unit overhead; it certainly requires both processing power and buffer storage resources at every node in the network, including the subscriber unit. The economic tradeoff is easy to state, hard to calculate:

1. If "lines" (including cellular) are cheap, use circuit switching.
2. If "computing" (including devices) is cheap, use packet switching.

It is not surprising that Europe often led the way in packet switching because of its historically punitive communication tariffs. The useful measurement of these competing costs in the United States is the subject of the next chapter.

REFERENCES

[1] *Mobile Data Report*, 1-15-90.
[2] *Mobile Data Report*, 1-14-91.
[3] *Mobile Data Report*, 5-6-91.
[4] *En Route Technology*, 7-8-92.
[5] *On the Air*, Spring 1993, p. 5.
[6] *On the Air*, Winter 1993, p. 4.
[7] *Mobile Data Report*, 9-9-91.

PART

BUSINESS

CHAPTER

3

FITTING APPLICATIONS TO PUBLIC OFFERINGS

3.1 NETWORK POSITIONING

We are always interested in simple ways to classify complex choices. An initial "application fit" technique for terrestrial, two-way wireless data networks is shown in Figure 3-1.

3.2 MESSAGE LENGTH VERSUS RATE

If traffic activity is brisk but message lengths are short, public packet switched networks are a reasonable business choice. If traffic activity is modest but message lengths are long, transaction-dedicated cellular channels are a better fit. This category includes "pure" (image) facsimile. If the application requires both high activity and long messages, a private network is

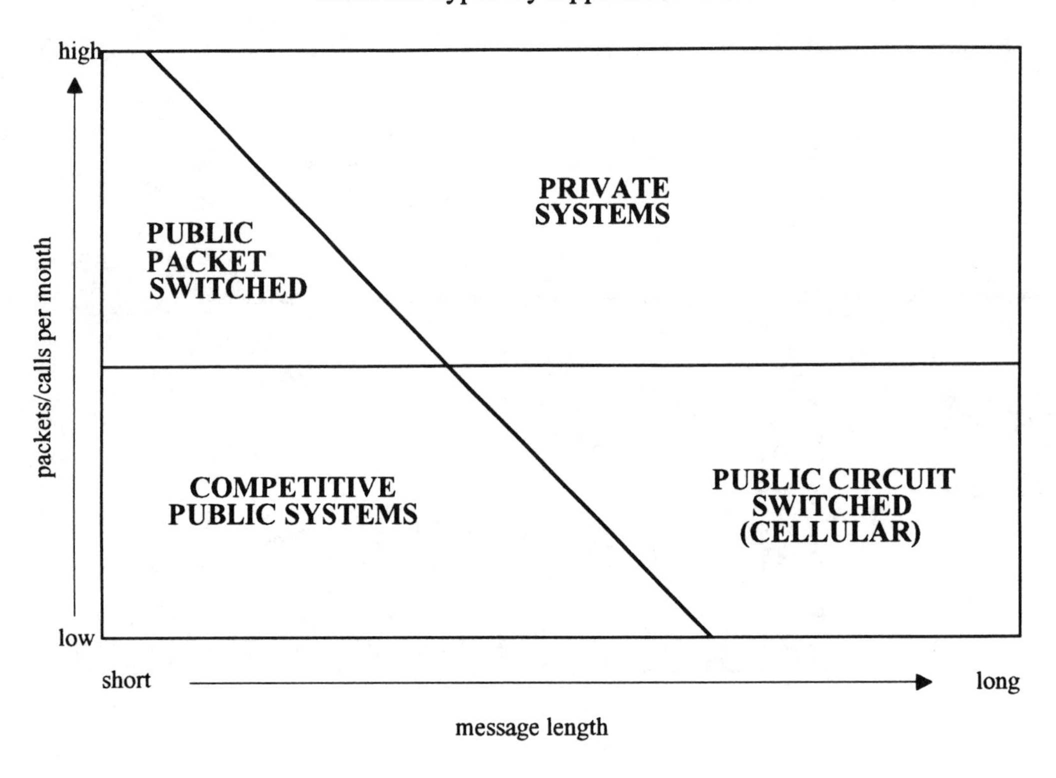

Figure 3-1 *Network Types by Application Class*

appropriate. Public packet switched and cellular systems can be extremely price competitive for application profiles featuring both medium activity and message length.

3.3 INITIAL PRICE POSITIONING: CURRENT NETWORKS

3.3.1 Public Packet Switched Networks

Reasonable estimates of price sensitivity to both frequency and length are possible, at least for the list price (both major service providers have bulk rate pricing).[1] One needs only to specify the target monthly fee, which, for illustration purposes, is set at $75.

RAM's packet charges[2] can be represented as $.025 for the first 12 octets, the natural size of the Mobitex primary block, and $.01 for every additional 50 octets, resetting at the maximum packet size of 512 octets. There is a $30 per month subscriber charge for wide area coverage.

ARDIS's standard rate[3] is $.08 per empty packet (maximum size: 240 octets), plus $.0004 per character transmitted. ARDIS has a "country club" minimum monthly charge of $32. If the subscriber unit is idle (vacation, etc.) the $32 must be paid. But if the user is active, the $32 is written off against character charges.

The message frequency versus activity comparison is shown in Figure 3-2. The choice of a log/log scale is used to illustrate the capability of the public packet switched networks to handle high-rate/short-length messages.

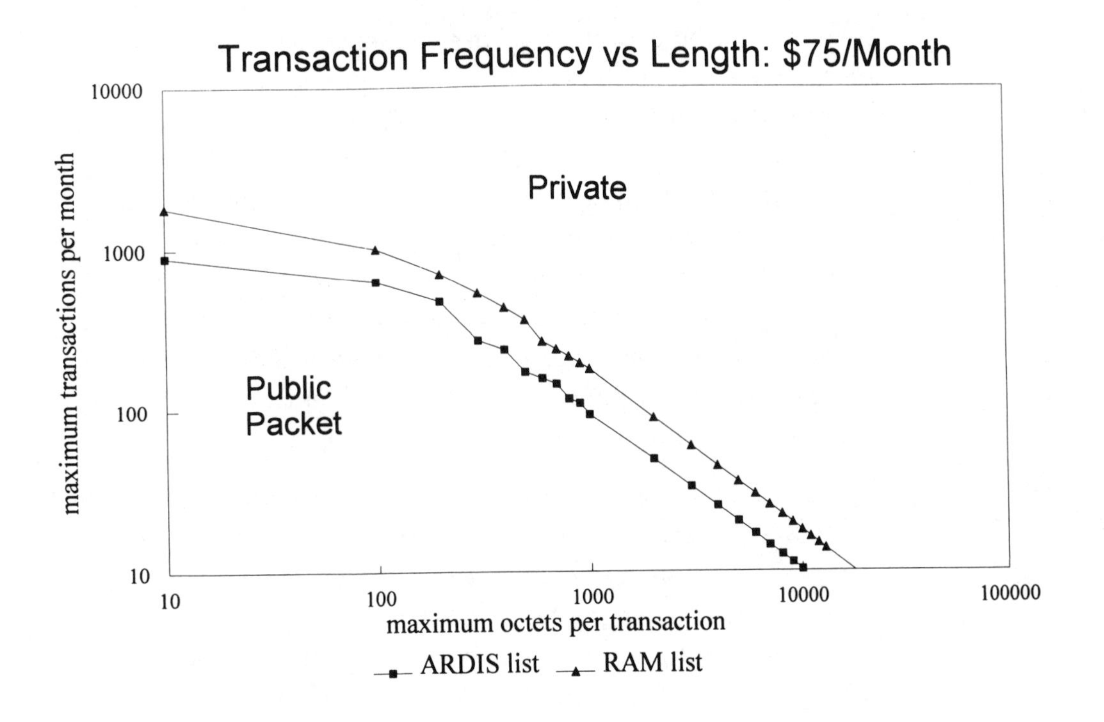

Figure 3-2 *Transaction Frequency Versus Length: ARDIS and RAM List*

Figure 3-2 illustrates that RAM is significantly more price competitive than ARDIS. This is probably natural given the exceedingly low number of subscribers currently using RAM (see Chapter 5). ARDIS simply has not yet felt serious competitive pressure from RAM.

However, RAM has just completed its nationwide infrastructure rollout[4] of 840 base stations covering 210 metropolitan areas. Disagreements exist between the two service providers as to the degree of coverage each now offers in these zones of high business activity. ARDIS appears to have a modest edge in base station count: 925 to 840; there is virtual equivalence in the number of base station sites.[5] ARDIS likely has superior in-building penetration; but objective comparitive measurements have not been completed. RAM likely has broader coverage. RAM should begin posing serious competition to ARDIS in 1993.

ARDIS can be expected to correct its pricing structure to be RAM-competitive. Indeed, it already provides the same air-time pricing for RadioMail as RAM.[6] The natural granularity of the new ARDIS protocol (RD-LAP) is 12 octets, which smoothly matches both the 6-octet granularity and 240-octet maximum packet size of the original protocol (MDC4800). Short ARDIS messages have low retry rates but high packet overhead. Longer messages spread the overhead more efficiently but retry more often in order to achieve clean delivery of the packet. The actual cost per clean user bit tends to be remarkably constant across the full range of the packet.

A simple business stroke would be to adjust packet prices to $.0128 per 12 octets—period. An infinite number of other choices are possible, of course. But at this level the packet is revenue neutral at length 120 (the approximate ARDIS average message length), more expensive at longer lengths, and more competitive for short lengths. The result is shown in Figure 3-3 and indicates how fiercely competitive the two public offerings are likely to be at the 100-octet (or less) mark.

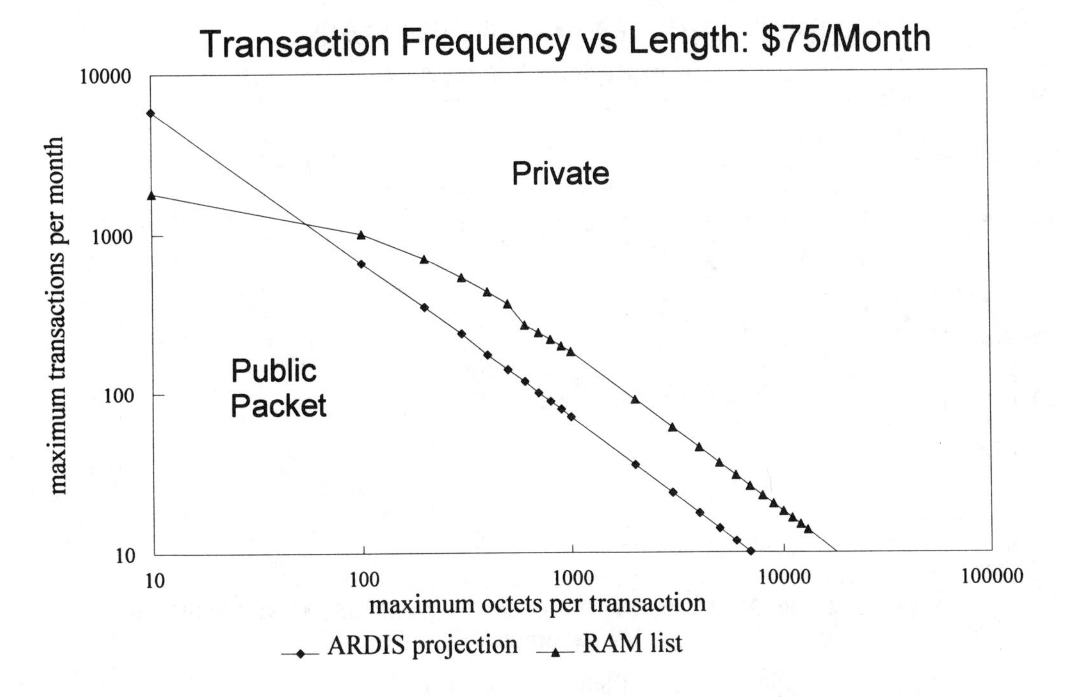

Figure 3-3 *Transaction Frequency Versus Length: ARDIS Projection Versus RAM List*

This smooth ARDIS price projection will be used as a surrogate for comparison to other existing alternatives.

3.3.2 Data-over-Cellular

The first step is to select a representative cellular tariff. In the Northeast, Metro Mobile offers ten prime-time plans.[7] The extremes were eliminated: corporate multiphone discounts as well as variations of consumer "Comfort Plans." This yielded three comparable price plans for the $75 per month price target: professional plan N3, business plan N5, and standard plan B. As shown in Figure 3-4, the resulting price is very close for 75–110 monthly minutes (and the $75 target).

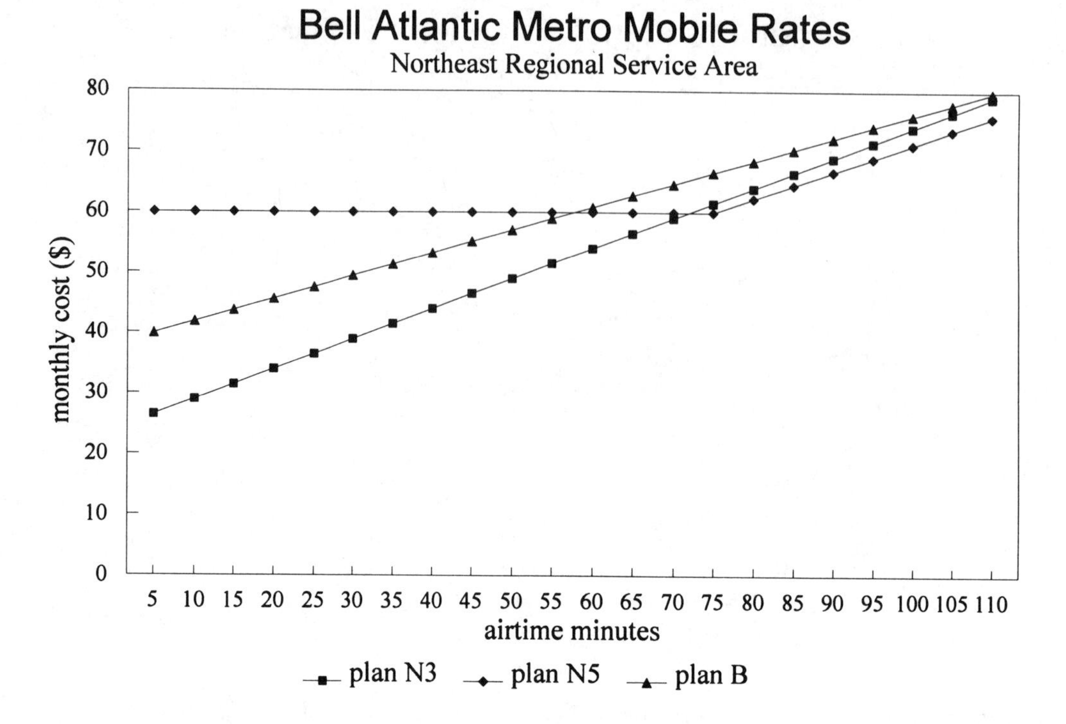

Figure 3-4 *Bell Atlantic Metro Mobile Rates*

Next, the decision must be made as to whether the cellular phone exists for data only—thus absorbing all of the fixed monthly fee—or would exist for voice purposes anyway so that the data costs are only incremental minutes. For this comparison the standard plan was selected, with no fixed-fee burden applied to data. The result: $.38 per minute for data.

Cellular data-minute consumption will vary because of the differences in modem speeds, which influence the raw transmission time as well as the employment (or need) for forward error detection/correction (see Chapter 9). In general, one wants to transmit as rapidly as possible to shorten connect time. Modem speeds continue to rise, but 14,400 bps fax/data modems rarely (if ever!) achieve such speeds via cellular. Two initial comparisons are made:

1. 4800 bps, with 25% of that bit rate lost to packet overhead (a data surrogate)
2. 7200 bps, with no overhead (a facsimile surrogate)

More annoying (and costly) is the long call setup time. As demonstrated in Appendix C the dial portion alone consumes ~9 (CompuServe) to ~13 (NewsNet) seconds via wireline. With cellular these times can easily double. The ID/password process can be automated, but the information providers seem unable to resist advertising their new wares (NewsNet can be set to advertise only once per day). Further, initial menu selections must be provided. Similar handshaking occurs with facsimile call establishment. The first 30 seconds of every cellular call are considered wasted in this analysis.

The results are examined in Figure 3-5. A one-minute call (30 seconds useful) permits about 200 transactions per month at the

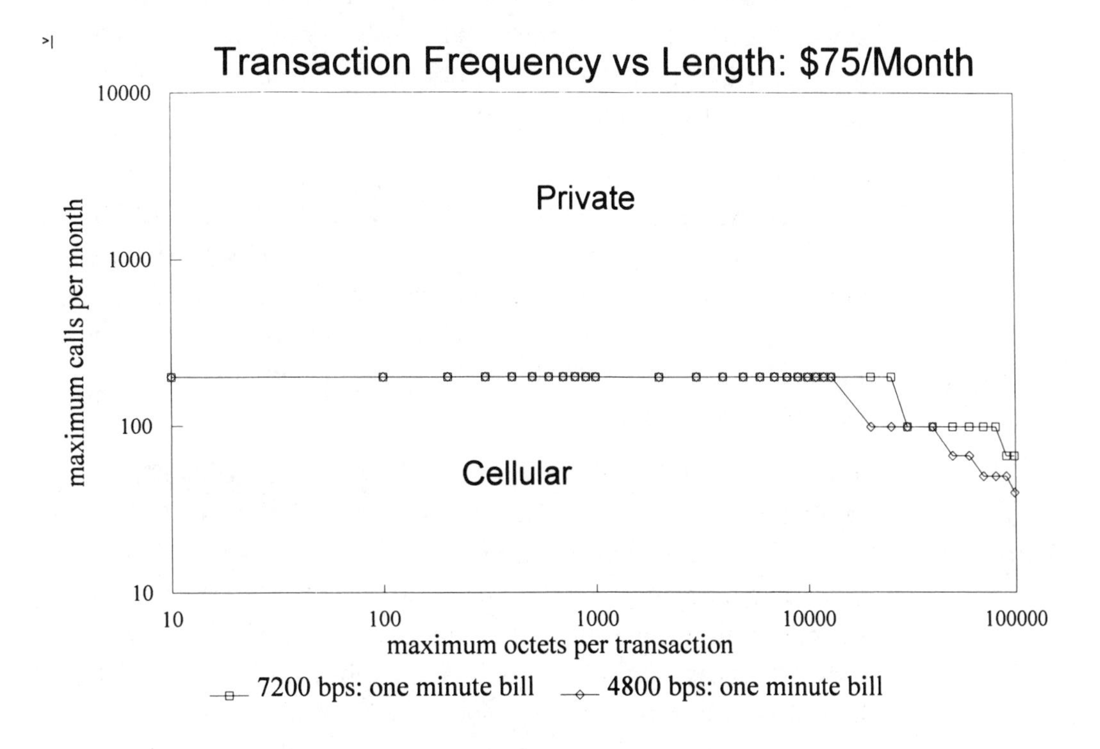

Figure 3-5 *Transaction Frequency Versus Length: Current Cellular*

$75 target. Note that the increased transmission speeds barely improve the message-length characteristic because:

1. of the setup time penalty,
2. current one-minute tariffs are punitive for subminute traffic.

The higher speeds *are* valuable for fax users. A full single-page, high-resolution facsimile image is about 200,000 octets after compression. Thus, about 200 such single-page faxes can be sent each month and stay within the $75 target. This probably meets the needs of such users as Sierra Semiconductor's sales manager, who "must send 10 to 15 faxes a day from my PC."[8]

Whereas setup time overhead is only reluctantly beginning to yield to new engineering approaches, the cellular carriers could correct the full-minute billing penalty at once. Clearly the cellular carriers should adopt one-second billing were data the only consideration. However, the impact of lost voice revenue because of a granular data tariff puts thoughtful business pressure on efforts to "tune the tariff."

This is not a new problem. A key selling point of the early SPAN modems[9] was their location at a value-added network (VAN) or MTSO, accessible via a special number that clearly distinguished a voice call from data. Special product solutions also exist to permit the carrier's switch to distinguish a voice call from data/fax calls: MPR Teltech offerings to Bell Cellular[10] is probably best known.

Bell Atlantic Mobile is now testing the Canadian philosophy with a view toward a special data tariff in 1993.[11] PacTel Cellular has been considering subminute data billing since 1992.[12] AT&T has filed an international fax tariff that, after the first 30 seconds, bills in 6-second increments.[13] However, Ameritech with its Chicago "modem pool" service has declined to offer data pricing because "the rates are low enough already. Most customers have some type of corporate plan that provides them with . . . peak time rates . . . less than 30 cents per minute."[14]

Cellular data tariffs will obviously become more attractive in response to market opportunities. The impact of one-second billing is shown in Figure 3-6. At low (100–500 octets) message lengths the number of supportable messages jumps to nearly 400 per month. At 7200 bps nearly 230 single-page facsimiles per month can be transmitted.

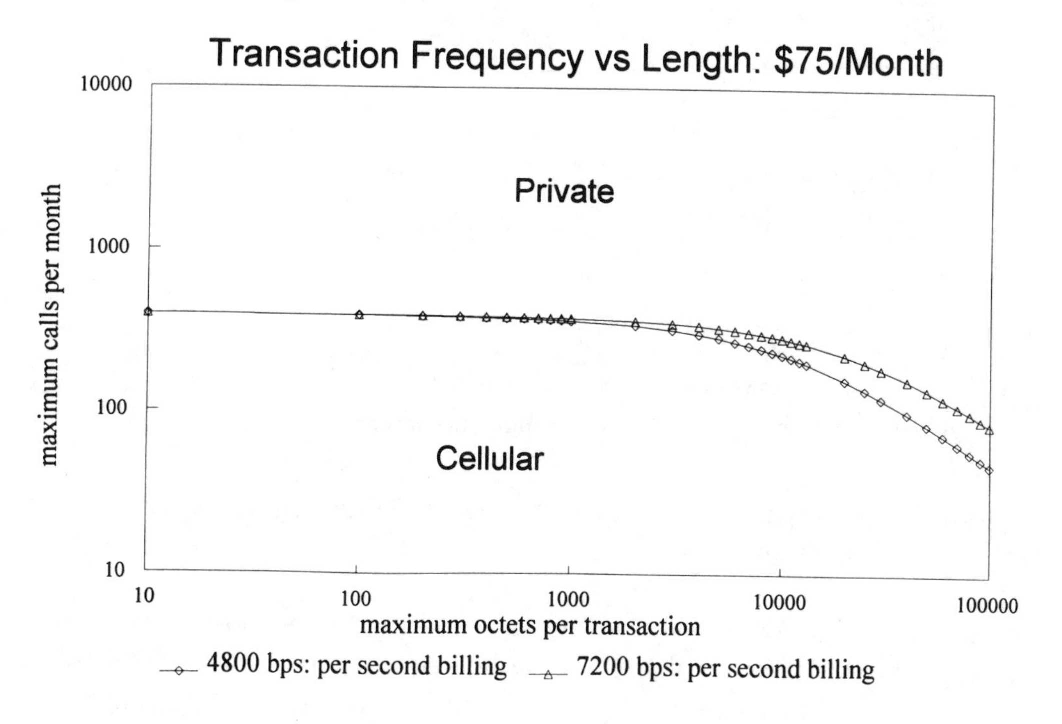

Figure 3-6 *Transaction Frequency Versus Length: Per-Second Cellular Billing*

The smooth 4800-bps-per-second curve is used as a surrogate for data-over-cellular.

3.3.3 Public Packet Switched versus Data-over-Cellular: Price Comparisons

A simplified price overlay of packet switched versus cellular (Figure 3-7) reveals the clearest strengths of each approach.

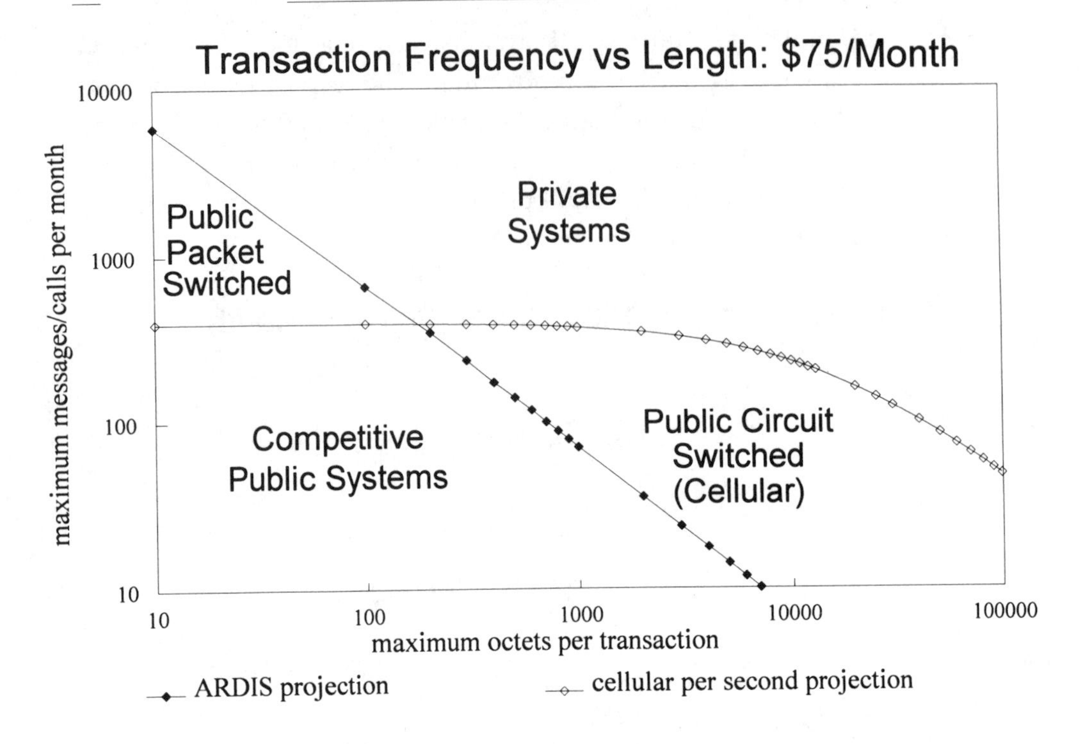

Figure 3-7 *Transaction Frequency Versus Length: Packet Switched Versus Cellular*

The ARDIS price projection was chosen to demonstrate packet switched capability with short, frequent messages; the 4800 bps cellular one-second billing rate portrays the medium-length, medium-rate message strength of cellular. The theoretical crossover point is a single packet ~200 octets in length occurring ~390 times per month.

Because the best-case cellular rate is determined by dial/setup time, even with one-second billing a message frequency higher than 390 calls per month (~20 per working day) cannot be contained in the $75/month constraint. Below that critical rate cellular begins to compete when user message lengths exceed 200 octets—the longer the better.

The log/log scale permits one to focus on the area in which the two solutions compete fiercely. The pricing structure of both approaches can often be tuned so the selection decision must be made on secondary application characteristics: the time between successive inquiries (short "think time" favors cellular), inability to tolerate dial delay (favors packet switched), requirement to have voice backup (favors cellular), and so on.

3.4 INTERACTIVE TRANSACTIONS AND THE IMPACT OF "THINK TIME"

Data-over-cellular sometimes can have advantages over packet switched if the application is interactive (database inquiry, some types of e-mail). A single inquiry/response counts as *two* packet switched transactions. The same inquiry/response can be completed in a single cellular call, but system response time consumes chargeable airtime.

To illustrate this point, assume:

1. Current RAM charges:
   —Subscription fee: $30 per month
   —Packet fee (max. 512 octets per packet):
      —$.025 for 1st 12 octets
      —$.01 for each additional 50 octets
2. Cellular charges:
   —Subscription fee: charged against voice
   —Time charge: $.38 per minute
   —Full-minute billing
3. An interactive application with 40 octets in/800 octets out per transaction
4. Four transactions per session
5. Fifty sessions per month
6. First 30 seconds of cellular time unproductive
7. Think time between transactions of one minute

Monthly RAM costs for this application type are:

Subs	Packet charges Inbound	Outbound		Session		Month		Total
\$30 +	((\$.035) +	(.125 + .085))	× 4		×	50	=	\$79

Monthly cellular costs for this application type are approximately the same, \$76: 50 sessions at \$1.52 each (Figure 3-8).

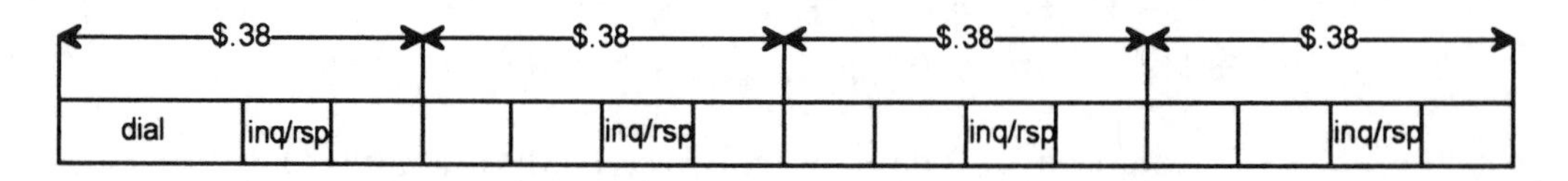

Figure 3-8 *Cellular Cost Impact: 60-Second "Think Time"*

Note that the first 30 seconds are consumed by dial time and overhead and the final 15 seconds are wasted.

However, if the think time between transactions is 45 seconds (Figure 3-9), typical for CICS applications,[15] the cellular session cost falls to \$1.14 and the monthly bill to \$57, illustrating the importance of exact application knowledge.

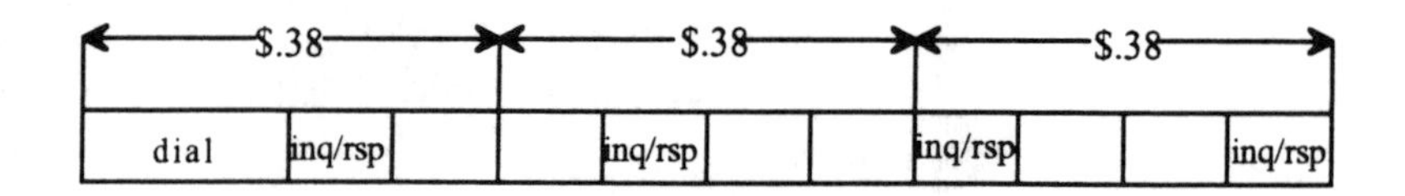

Figure 3-9 *Cellular Cost Impact: 45-Second "Think Time"*

This example also illustrates that cellular techniques that shorten the interval between successive transmissions, from improved modem bit rates to data compression, may quickly pay off in reduced airtime.

3.5 MORE CONFUSION: THE "FAX SERVER"

Several prior pages dealt with the transmission of facsimile images via cellular. This is very often confused with RAM's announcement of its intent to provide "fax servers" in 1993.[16]

The RAM service will not be bit-image transmission. Instead, RAM will transmit normal data or text messages (which are a small fraction of the length of a bit image) to a gateway processor at the subscriber's host complex. The gateway will convert the message to the appropriate typeface, merge it with a prestored background page such as a letterhead, and add requested information. The now completed, very much larger, desktop publishing–quality document is then sent as a true facsimile over landlines. After successful fax transmission a short confirmation is sent to the subscriber unit via RAM. The packet switched wireless transmission task thus consists of normal-length messages, not a huge facsimile image.

REFERENCES

[1] Telemeter Corp (wireless utility metering) is one RAM example; see *En Route Technology*, 6-7-93.

[2] John Krachenfels, RAM Director of Business Development, as reported in *Industrial Communications*, 9-28-90; and William Frezza, Ericsson GE Director of Strategic Sales, as reported in *Communications Week*, 7-20-92.

[3] *ARDIS Pricing Sheet*, 10-8-93 (still valid 7-27-93).

[4] *Telecommunications Alert*, 6-23-93.

[5] *Mobile Data Report*, 7-5-93.

[6] *Mobile Data Report*, 8-30-93.

[7] Bell Atlantic *Metro Mobile Price Plans*, 10/92 (still valid 9-4-93).

[8] Doug Rasor, Sierra Sales Manager, *Mobile Office*, April 1993, p. 34.

[9] Spectrum Cellular & CompuServe agreement as reported in *Mobile Phone News*, 2-12-86; and Spectrum Cellular & Ameritech agreement as reported in *Communications Week*, 3-2-87.

[10] As reported in *Mobile Data Report*, 12-16-91.

[11] Benjamin L. Scott, EVP and COO, Bell Atlantic Mobile, as reported in *Telephone Week*, 6-21-93.

[12] "If we can't do sub-minute billing, we're not going to have a (data) market." Lee Franklin, President, PacTel Wireless Data Division, as reported in *Mobile Data Report*, 11-23-92.

[13] *Telecommunications Reports International*, 6-11-93.

[14] Greg Oslan, Ameritech Marketing Manager, Cellular Data Services Group, *Mobile Data Report*, 8-30-93.

[15] IBM *System Selection Guide, ZZ20-4318-13,* CICS with MVS/XA Workload Description, p. 80–60.

[16] *New York Times,* 11-8-92.

CHAPTER

4

PUBLIC NETWORK ALTERNATIVES

4.1 USING DATA-OVER-CELLULAR

The first cellular system became operational in October 1983. Nearly simultaneously Tandy introduced its first notebook computer, the TRS-80 Model 100. Users quickly realized that Tandy's cellular phone "portable adapter kit" permitted an awkward—but workable—physical connection of the notebook to the cellular network. Using the TRS-80's slow (300 bps) internal modem, and a great deal of manual intervention, journalists began to send publication copy from the story site itself, though not while in motion. A market opportunity was detected.

4.1.1 Proprietary Error-Correction Devices

In March 1985 Spectrum Cellular introduced its first 300 bps cellular modem with forward error correction; the speed was increased to 1200 bps in April 1986[1] and reached 2400 bps in 1989.[2] The Spectrum approach featured a proprietary protocol that was installed in a specific modem pair: the Span and the Bridge.

Both units were virtually identical in external appearance. The videocassette-size Span was intended for "fixed-base" equipment such as the communications front end of mainframes. Its physical connection was to conventional wireline services. The Bridge was attached to the mobile unit. Because it had no radio of its own it was required to attach to a reasonably rich mix of cellular phones to obtain the radio function. A representative configuration is shown in Figure 4-1.

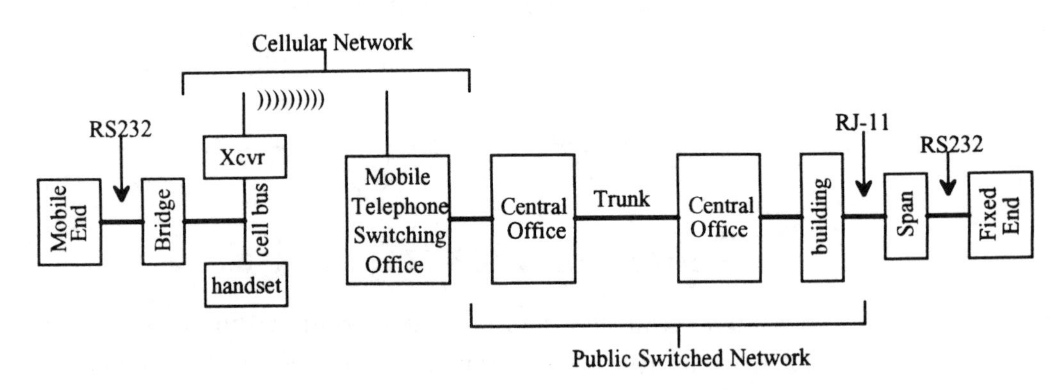

Figure 4-1 *Cellular Device to Fixed-End Connection: Spectrum Approach*

A practical business problem of the Bridge/Span combination for private systems was the need to enlist the cooperation of the data processing manager to install Span modems at the fixed site. Value-added networks (VANs), such as CompuServe, initially responded positively; five cities were operational by February 1986.[3]

But a noteworthy variation quickly evolved. Interested cellular carriers moved the Span forward to the mobile telephone switching office (MTSO), where it was connected to a normal landline modem (see Figure 4-2).

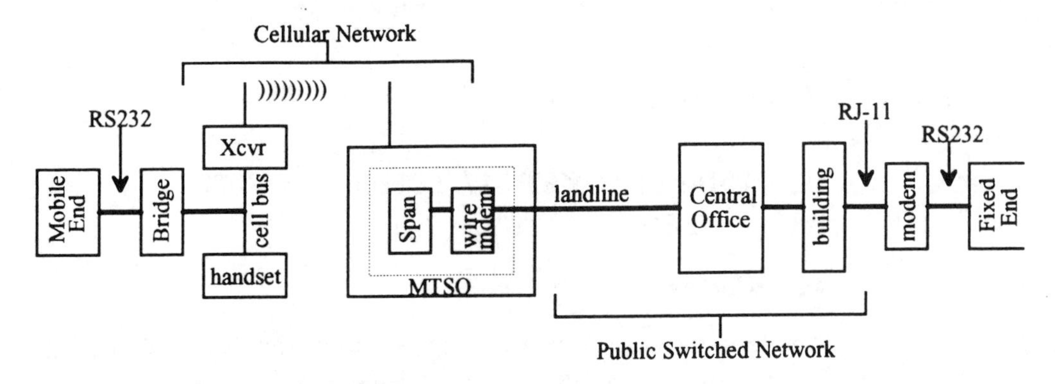

Figure 4-2 *Movement of Fixed-End Modems to MTSO*

The mobile user was thus free of the need to have a Span purchased and installed at the fixed end. The cellular service provider was able to offer a data-only service via designated numbers. By 1989, 43 "SPCL Data Service" markets were operational in the United States,[4] with 54 service providers. But the use of "311" prefixes to identify and place a premium on data calls, as Cantel did in Canada,[5] was not successful. By mid-1990 only ~12,000 Spectrum modems of all types had been sold in North America.[6]

The notion of direct connection to the MTSO was revived when UPS connected its UPSnet to the MTSOs of four cooperating cellular carriers in order to bypass the public landline network.[7] With this technique, and others such as short-burst traffic, UPS secured for itself exceptionally low cellular tariffs. The result: 50,000 data-only delivery information and acquisition devices (DIADs) installed in 1992—the largest single data installation in the world (see Figure 4-3).

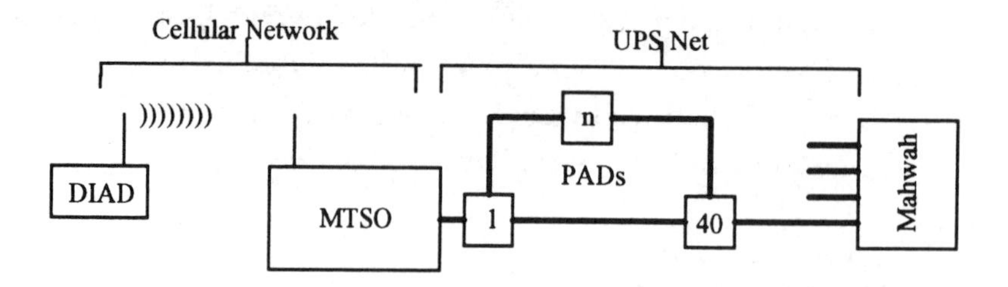

Figure 4-3 *UPD MTSO Connection*

4.2 PUBLIC PACKET SWITCHED SERVICE PROVIDERS

4.2.1 Categories

Amateur packet radio has been active in the 2- and 20-meter bands for years. But recognizable pay-for-service packet switched offerings began in Chicago in 1986 with Motorola's DRN. Current North American public packet networks can be organized into four categories:

1. Operational now
2. Field testing now
3. Planning stages
4. Terminated or otherwise inactive

The operational category has been further subdivided into pure data and hybrids, which often can be cost-justified solely on vehicle location (AVL) grounds. Because of the dominant position of satellite systems in this category, an exception has been made to the terrestrial-only format.

4.2.2 Operational Now

4.2.2.1 Data Only

Table 4-1 is a comparison of the two major competing data-only public packet switched services organized to highlight their key business characteristics. A detailed breakdown of the subscriber count summaries is available in Chapter 5.

Table 4-1 *Operational: Business Characteristics*

Provider	ARDIS		RAM
Ownership	IBM/Motorola		RAM/Bell South
Infra-structure Supplier	Motorola		Ericsson
Protocol	MDC4800	RD-LAP	Mobitex
Services, Applications:			
Principal emphasis	handheld devices: –in bldg penetration –short (~120) messages –low velocity users vertical applications no voice capability no image fax capability no AVL capability roaming w/new devices	handheld/vehicular: –in bldg penetration –medium (<512) msgs –metro speed users vertical applications no voice capability no image fax capability no AVL capability auto roaming support	portable/vehicle units: –street level coverage –short, long messages –metro speed users horizontal applications voice capable; no license no image fax capability AVL: zone location only roaming standard
Current coverage	~1300 base stations, 400 metro areas, 8000 towns 50 States, PR, VI (US); 9 Canadian Prov's; links: UK, HongKong, Grmny, Australia	~150 base stations:Wash NYC, SFO, LAX, 4 addl 1993; 2 protocol modem allows MDC4800 use ~900 base stations: 1996	~840 base stations, ~100 metro areas (USA); links to Mobitex Canada
Current NYC spectrum	250KHz	50KHz	500KHz
US subscribers	YE'90: ~22,000 YE'91: ~27,000 YE'92: ~31,000 1Q93: ~32,000		YE'90: 0 YE'91: <50 YE'92: <1000 1Q93: >1000
Pricing	minimum monthly charge: $32 –covers 1st $32 packet charges peak (7AM-6PM) access fees: –$.08 empty packet –$.0004/octet		monthly access: $15-30 peak (per 512 octets): –$.025/1st 12 octets –$.01/add'l 50 octets

Table 4-1 *Operational: Business Characteristics (continued)*

Provider	ARDIS		RAM
Subscriber Equipment:			
Radio packet modems	RPM 840C: external PC connectivity modem. 18 ounces; list $1650 InfoTAC: personal data communicator; 17.7 oz 4 protocols, list $1395; ARDIS price: $995 MRM420 (Mobile): multi-protocol; high transmit power RPM400i integrated; list $1500; qty $950=>535 RPM405i integrated; qty: $1000 = >595; small footprint, battery drain RPM415i integrated: Toshiba laptops		Mobidem 1090 external PC connectivity modem 16 ounces; list $775 Mobidem AT (Intel) e-mail unit; $745 Motorola RPM400i: Mobitex version (4Q92) Motorola RPM405i: Mobitex version (4Q92)
Sample integrated users	laptops: –AT&T Safari (dock) –IBM PCRadio –NEC 3125 (pen base) –Toshiba (multiple) handhelds: –Itron T5000 – Motorola KDTs –Poqet (Fujitsu) –Psion HC PDT220 –Telxon PTC860		laptops: –NCR 3170 handhelds: –Melard Access PC
Vehicular devices	KDT 9100-386		MTT40/C700 radio
Facsimile machines	N/A		N/A

Although each offering is distinct, and may optimize for a particular application, ARDIS and RAM are closely comparable. Neither has voice or "pure" facsimile capability, and depend only upon data for business success. Both are metro systems geared for national coverage. Both have ample near-term subscriber capacity. Neither are yet profitable, though ARDIS airtime revenue will probably exceed $6 million/month by the end of 1993.

ARDIS started earlier, has superior in-building penetration in major metro areas, reaches more small cities and towns, has a reasonable stable of field service applications, and is particularly strong in handheld, low-user-velocity applications. Its second-generation infrastructure, using an improved protocol, was operational in Washington, D.C., in October 1992. Eight cities are scheduled to be operational on the new protocol in 1993.[8]

RAM started later and spent much of the past 3.5 years installing infrastructure in major cities. By its own admission[9] its in-building penetration does not match ARDIS; it is possible that its cellular-like approach will always tend to preclude this equivalence. RAM concentrates on the top 100 or so major metropolitan areas, where its base-station counts are roughly comparable to ARDIS. It has made a series of arresting business partnership arrangements. Its prices are low. Attractive modem alternatives exist. RAM's subscriber track record has been painfully weak, but the service should be poised for growth in 1993.

Table 4-2 compares the salient *technical* characteristics of the two operational networks. Many of the summary points are discussed in detail in later chapters.

Table 4-2 *Operational: Technical Characteristics*

Provider	ARDIS		Mobitex USA
Network Connections:			
E-mail	RadioMail (thus Internet); RF Data		cc:Mail; DEC All-in-1; RadioMail (Intrnet); PSI
VAN	Advantis; BT/Tymnet; SprintNET		SprintNET (testing)
Private line	SNA 3270; SNA LU6.2; X.25; Async; BSC3270		X.25 LAP-B; BSC; TCP/IP
Airlink:			
Control channel	none required		Local System Channel minimum; NSC to roam
Channel: **bps** **width (KHz)** **Bits/hertz**	4800 25 .19	19,200 25 .77	8000 12.5 .64
Protocol efficiency	125 octet msg: ~36% 240 octet msg: ~39%	125 octet msg: ~57% 240 octet msg: ~61%	125 octet msg: ~51% 250 octet msg: ~55%
Random error strategy	convolutional 1/2, k=7; always fix 2 per 112 bits	trellis coded modulation; rate = 3/4	12,8 Hamming; always fix 1 per 12 bits
Burst error strategy	interleave 16 bits; withstand 3.3ms fade; retry complete message	interleave 32 bits; withstand 1.7 ms fade; retry complete message	interleave 21 bits; withstand 2.6ms fade; retry failing block(s)
UBER	CCITT CRC-16 limited; published: 1 in 100,000	CCITT CRC-32 limited; published: 1.4×10^{-11}	CCITT CRC-16 limited; 16 of 10,000 blocks with burst >16 bits will fail
90% success, 1st try	msg length <84 octets	msg length <45 octets	
Channel-in access	contention; CSMA non-persistent	contention; slot CSMA; 5 ms turn-on-time radio	contention; slot CSMA with silence order
Efficiency in	G = 45%; S = 30%	G= 80%; S = 40%	not simulated by JFD
In msg./sec., single chan-nel, base sta-tion (ideal)	length 50: .82 msg/sec length 100: .50 msg/sec length 150: .35 msg/sec	length 50: 4.66 msg/sec length 100: 3.29 msg/sec length 150: 2.48 msg/sec	

Table 4-2 *Operational: Technical Characteristics (continued)*

Provider	ARDIS		RAM
System Considerations:			
User information	only EBCDIC bytes and padded ASCII	transparent octets	transparent octets
	variable length packets; 6 octet granularity	variable length packets; 12 octet granularity	variable length packets; 12/18 octet granularity
	240 octet max packet; no adaptive assembly	512 octet max packet; 4096 octet max message	512 octet max message
Control philosophy	highly centralized metro area mgmt; low function base stations/devices		decentralized; high function base station(s)
Retry control	message CRC	multiple CRCs: hdr/msg	block CRC
	stop and wait ARQ		go-back-n ARQ
	separate header & ACK		separate header & ACK
System availability	99.97% 1992/1H93		

It is sufficient to note that the current ARDIS airtime protocol (MDC4800) is near end-of-life. It was first developed in 1980 for public-safety applications and became codified for IBM's DCS in 1983 when it was submitted to the IEEE 802.6 as a proposed metropolitan-area network standard. Although the bit rates and protocol efficiency of MDC4800 are no longer in the van, the adjacent cell frequency reuse approach remains unique and powerful. This approach is an option in the radio data link access procedure (RD-LAP) and is being deployed in that manner by ARDIS to improve in-building penetration.

The 8000 bps, 12.5 KHz wide, multiple-channel Mobitex implementation is technically superior to MDC4800 when deep building penetration is not the primary motivation. But, in its turn, it has been overtaken by the Motorola RD-LAP approach.

Analog technology improvements continue with CDPD. That protocol's bit rate is identical to RD-LAP, and its bits/hertz rating is the same as Mobitex. But it has a powerful array of other techniques that promise excellent future subscriber capacity. Meanwhile, Nextel introduced Motorola's digital MIRS technology in August 1993 in Los Angeles. Using linear technology, not frequency modulation (FM), 64 Kbps will be transmitted in a 25 KHz channel.

4.2.2.2 Hybrid: Data, Telemetry, AVL, Facsimile

CoveragePLUS is a very different system from ARDIS/RAM albeit sharing some of the same Motorola technology as ARDIS. It serves a hybrid market grounded in elementary automatic vehicle location (AVL); sophisticated AVL is available at extra cost via GPS. Thus its closest (and far more successful) competitor is satellite-based Omnitracs.

CoveragePLUS, first operational in 1989, is a linked, terrestrial special mobile radio (SMR) network. Dispatch voice is standard. A subscriber can be usefully operational with no data or associated application programs. Data is an optional add-on. Facsimile is available on voice channels using Motorola units. There is no real metro coverage, although Motorola has purchased a minority stake in RaCoTek[10] with a view toward CoveragePLUS/RaCoNet linkage. Basic device prices begin at ~$1,850.

The RaCoNet system is also based on connecting independent SMRs. It is "connectionless . . . (and) interactive communications are not possible. Response times are typically several seconds, and dead times" are common.[11] RaCoNet began beta testing with five companies in October 1990[12] on E. F. Johnson SMRs. Subscriber progress has been slow. After 18 months of effort there were only "several commercial installations"[13]; after 2.5 years there were "slightly fewer than 100 customers"[14];

in August 1993 there were conflicting reports of customers ranging as low as 70[15] to a high of 100.[16] Airtime prices ($.02 per 100 octets)[17] are competitive, but device prices are higher than CoveragePLUS: $3,500–$5,000 (including radio).[18]

Cooperative business agreements were signed with Motorola,[19] Nextel (formerly FleetCall),[20] and Advanced Mobilecomm (New England)[21] in 1992. In 1993 Datatrac Corp.[22] began a partnership with RaCoNet for its courier order processing system (COPS), to provide voice and data for delivery drivers; ADAQ Systems began offering its microEXPRESS software on RaCoNet for package delivery companies[23]; Data Over Radio System[24] has a RaCoNet-specific application software package; and Arrowsmith Technologies[25] became a partner with its Fleetcon System for the cable television industry.

Qualcomm's Omnitracs system, first operational in 1988, is a two-way satellite (GTE SpaceNET Gstar I) based offering. AVL is standard, and two-way messaging and vehicle sensor telemetry are extra cost features. Neither voice nor facsimile is possible given the low bit rates to individual devices. Coverage is "rural" (highway) and pervasive—including Europe and Latin America. Basic device prices begin at ~$4,000.

Table 4-3 summarizes the key characteristics of these hybrid systems.

Table 4-3 *Operational: Summary Characteristics*

Provider	Omnitracs	CoveragePLUS	RaCoNet
Ownership	Qualcomm	Motorola	RaCoTek/Motorola
Infrastructure supplier	N/A	Motorola	EFJohnson/Motorola
Services, Applications:			
Principal emphasis	vehicle units: –highway coverage, –long (2000) messages no voice no facsimile AVL: QASPAR/GPS	vehicle units: –highway coverage –short messages dispatch voice stndrd; PSTN voice optional (weak) optional fax zone AVL; GPS option	vehicular solutions: –metro area street level –short messages dispatch voice via SMR N/A via EFJohnson N/A
Current coverage	non-metro USA and Europe	~600 SMRs operational, USA wide	spotty
US subscribers	YE'89: ~6500 YE'90: ~11500 YE'91: ~22000 YE'92: ~38000 2Q93: ~45000	YE'90: ~1500 YE'91: ~2300 YE'92: ~2900 2Q93: ~3300	YE'92:<1000 3Q93: ~2000
Economics:			
Pricing structure	monthly access: $35 (includes 800 messages) additional messages: –$.15 + $.002/octet	monthly access: $35 (incl. location/status); prime time: –$.05/240 octets	– $.02/100 octets
Timing:			
		announce to 3 cities: 1yr < 60 mile holes: 2.5 yrs	beta test start: 10/90
Subscriber Equipment:			
Vehicular devices	Omnitracs IBM Roadrider	MDT 7100/MoStar radio	DOS based PCs, with RaCoTek modems
Facsimile machines	N/A	MFax" 4" × 4" × 11.5"	
Network/Airlink Connections:			
VAN		BT/Tymnet	
Private line	AS/400	AS/400; Macintosh	

4.2.3 Testing Now

Numerous experimental tests of particular applications are underway. But there are two major terrestrial candidates to track:

1. Cellular Digital Packet Data (CDPD), a data-only analog technique for use either on dedicated channels or sharing idle time on voice cellular channels. In the latter case voice has priority.
2. Nextel's ESMR system, a digital technique for both voice and data; the initial system, featuring digital paging displayed on the voice instrument, went online in Los Angeles in August 1993.[26] Facsimile, two-way data, and AVL are scheduled for later rollout.

CDPD has the endorsement of five RBOC mobile subsidiaries and two non-wireline carriers; many are committed to full-scale testing in 1993. McCaw has selected Las Vegas for its first commercial experimentation beginning in August (now slipped to October)[27] and is committed to bringing wireless data services to 50 percent of its national network by year's end and completing the CDPD rollout by the second quarter of 1994.[28]

Bell Atlantic Mobile is scheduled to begin its initial testing in Washington, D.C., in November 1993, with a plan to offer CDPD in markets "where commercially viable" at the end of 1994. Infrastructure will not be offered at all cell sites.[29] Ameritech has a fourth-quarter plan for Chicago using Hughes base stations.

Nextel employs Motorola's digital MIRS (Motorola Integrated Radio System) technology, which offers the promise of high (64 Kbps) data rates—and is enormously challenging to competition. LA Cellular (65 percent owned by Bell South Mobility and 35 percent by McCaw) has announced a discount pricing plan that tends to bind existing customers to it and prevent movement to Nextel. Nextel has appealed the new pricing plan to the California Public Utility Commission.[30] Meanwhile, Nextel leads "a consortium of eight major SMR operators toward a nationwide, uniform . . . enhanced SMR system."[31]

Finally, CDI's tests with Bell Atlantic/Westinghouse—a departure from the fixed-position, low-duty cycle applications originally tested by GTE Mobilnet and BAMS—are included because Bell Atlantic has invested in CDI and testing was still in progress in June 1993.[32]

Table 4-4 summarizes the salient business characteristics for these three solutions.

Table 4-4 *Testing: Business Characteristics*

Provider	CDPD	Nextel	Bell Atlantic Mobile
Infrastructure supplier	multiple vendors	Motorola	CDI
Services, Applications:			
Principal emphasis	laptop PCs: –cellular penetration –medium length msgs –metro speed users	dispatch data: –high bit rate potential	vehicular LTL solutions
		digital paging	
		digital voice standard	
	voice via cellular		voice via cellular
	facsimile via cellular	facsimile	facsimile via cellular
	no AVL capability	AVL	AVL claim: to cell site?
	designed for roaming	designed for roaming	
First commercial site	Las Vegas, NV 10/93 McCaw	Los Angeles 8/93 134 voice cells	Washington/Baltimore Balt: ~27 bases
Planned coverage	coalitions of carriers to duplicate cellular net nationwide	SFO Bay Area 4Q93 NYC, CHI 1994 DFW, Houston 1995 nationwide 1996	
Economics:			
Infrastructure invest	unknown	IPO (1/92) $112M Matsushita(2/92) 45M Motorola(3/92) 260M NoTelcom(3/92) 40M Comcast(9/92) 50M Comcast(7/93) 114M	cell cost: $10-50,000 ea. dvlpmnt funds: $12.2M –round 1 1.2M –round 2 4.5M –round 3 6.5M
Subscriber Equipment:			
Radio packet modems	Cincinnati Microwave	digital: none needed	RT2004 (200mW)
Handheld devices	none known	integrated digital pager	N/A
Vehicular devices	none known	integrated digital pager	N/A

Table 4-5 summarizes the key *technical* characteristics of the systems in test. As expected, a great deal is known about CDPD because its specifications are truly public domain. Identifying characteristics include:[33]

1. The ability to coexist with voice on existing cellular channels (voice has priority). Alternatively, a dedicated-channel, data-only system is possible.
2. New base stations that exactly map the cellular sites in a voice coexistence system, with necessary implications on cost and schedule. Adaptive power control is integral to the system design.
3. A very large active user capacity in voice coexistence systems. This is achieved through a series of skillful design techniques, especially full-duplex radios, a small collision window, and selective ARQ.
4. An optimum target speed expected to lie between 25 and 30 mph for single packet messages.
5. A forward error-correction technique optimized for protocol efficiency at the cost of a relatively high undetected block error rate. This technique also defines minimum block size and constrains the maximum bit rate.

Table 4-5 *Testing: Technical Characteristics*

Provider	CDPD	Nextel	Bell Atlantic Mobile
Philosophy	analog AMPS coexist	digital TDMA; 6 time slots/25KHz; frequency re-use; low power sites; goal=15x current capcty	interstitial transmission, low bit rates/ low power, to avoid interference with existing voice
Airlink:			
bps/channel **Channel width (KHz)** **Bits/hertz**	19,200 30 .64	64,000 25 2.56	2,400 3 .8
Inbound (best) protocol efficiency (no errors, retries...)	125 octets: 48% 250 octets: 54% 500 octets: 62%		
Random error strategy	cover with burst protect		
Burst error strategy	R/S 63,47; withstand 2.2 ms fade w/o static errors; then retry packet		
Undetected block errors	2.75×10^{-8} (17 dB C/I)		
Inbound access	contention: slotted CSMA/CD		
Inbound efficiency(s) (S): G=70%	50 octets: 39% 100 octets: 40% 150 octets: 40%		
Inbound msgs/sec., dedicated single channel, base station (ideal)	50 octets: 4.42 100 octets: 2.52 150 octets: 2.33		
System Considerations:			
User information	transparent octets. max packet ~128 octets multi-packet variable length messages		
Control information	R/S each 378 bits; selective ARQ		

4.2.4 Planning Stages

Just as there are numerous niche systems in test, several special systems are being planned. The FCC recently set aside spectrum for new *digital* wireless communications. This news sparked an announcement from Mobile Communications Technologies of Jackson, Mississippi, that they "will . . . provide services over a 300 market network."[34]

A 1991 planning example is Advanced Mobilecomm's Mobile Radio New England (MRNE), a street-level, vehicular system for digital SMR.[35] As its name implies, MRNE is a regional system with only 25 transmitter sites throughout New England.[36] Although network connections to the six other Advanced Mobilecomm systems are feasible, no plan yet exists that is compatible with the current, non-networked, E. F. Johnson base.

Another 1991 example was the partnership between Ameritech and Mobile Electronic Tracking Systems (METS). The announced goal[37] was to build a nationwide, public, wireless data network that will start in Chicago. Ameritech has now dissolved the partnership.[38] METS' plans are unclear.

Other (still regional) examples include:

1. Nynex Mobile's plan to test a Bison Data system in Buffalo, New York, for automated teller machines and point-of-sale devices.[39]
2. Geotek's analog SMR data system, available in ten locations, which will begin conversion to digital in mid-1994 using Gandalf modems (Geotek owns 66 percent of Gandalf Mobile) and E. F. Johnson and Ericsson GE radios.[40] Note that Geotek has nationwide ambitions.
3. Teletrac's announcement that it will upgrade its terrestrial AVL system to provide auto acknowledgment and then two-way text messaging.[41]

A summary of those planning (seemingly) nationwide systems is in Table 4-6.

Table 4-6 *Planning: Summary Characteristics*

Provider	GeoNet	NWN	Teletrac
Ownership	Geotek	MTEL	PacTel
Infrastructure supplier	unknown	Motorola	Teletrac
Services, Applications:			
Principal emphasis	turnkey digital voice and data for 10-200 vehicle vertical markets AVL, dispatch, 2-way messaging, wireless fax, E-mail...	messaging for palmtop & laptop PCs; 3000 octet expected msg length	short msgs (max: 98); canned responses; vehicle "Post-it-Notes" AVL standard (FCC license constraint)
Pilot locations	Philadelphia 6/94	Dallas (demo) 4/93	6 city: status msg 5/93 2-way text 4Q93
Planned coverage	50 markets by 1996	300 markets by 7/95; 3000 base stations; 2,000,000 user capacity	NYC to be added later
Pricing			est. average: $30/month (includes AVL)
Devices			Coded CMX-4500 $800 Coded CMX-1000 $600
Technical Approach:			
Philosophy	narrowband digital frequency hopping multiple access (FHMA); full duplex voice/data via single "handset"	3 nationwide 50 KHz channels; narrowband simulcast	5 watt spread spectrum; locate vehicle within 150 feet
Airlink Specification:			
bps/channel **Channel width (KHz)** **Bits/hertz**		24,000 50 .48	
Burst error strategy		"closest to Reed-Solomon"	

4.2.5 Terminated

For some systems the future is already behind them. In the United States, Millicom terminated its system in 1990; DRN (of Vienna, VA—not the Motorola offering with the same acronym that became part of ARDIS) ended its seven-year effort in August 1991.[42] Motorola's own Data over Privacy Plus ended its existence at the close of 1991,[43] extinguished by competition (some internal) and only a finite opportunity.

There were additional shutdowns in Canada and Europe, most notably the decision of Motorola to abandon its plans to operate a mobile data network in the United Kingdom.[44]

Table 4-7 lists the salient characteristics of these systems.

Table 4-7 *Terminated: Summary Characteristics*

Provider	PrivacyPlus	Millicom	DRN (Vienna, VA)
Ownership	Motorola SMRs	Millicom	20/80 Motorola/Ademco
Infrastructure supplier	Motorola	none found	Ademco
Services, Applications:			
Principal emphasis	SMR user whose radio is in group call <10% of time (voice blocks data); SAS field service pckge	SMR users with equal voice & data needs	credit card authorization
Planned coverage	metro SMRs	nationwide	nationwide: 70 cities had devices when terminated
Subscriber Equipment:			
Vehicular devices	KDT440 (num): $1600 KDT460 (a/n): $2200 MoStar radios		
Stationary devices			POS devices
Technical Goals:			
Philosophy	voice has priority; data retries at 1 minute intervals if voice present		
Airlink specification	MDC4800	LTR signal format	RLP (Citicorp prop'tary)
Timing:			
	Data/PP announced 1/89 CoveragePLUS 6/89 ARDIS 1/90 RaCoNet beta test 10/90 Data/PP ended 12/91	waiver granted 5/90 test begins 8/90 suspended 12/90	founded (Citicorp) 1984 bought by Ademco 1986 Motorola: 20% 5/89 discontinued 8/91

REFERENCES

[1] Lotus, April 1986 edition, p. 27.
[2] *Mobile Phone News*, 12-21-89.
[3] *Mobile Phone News*, 12-12-86.
[4] SPCL Data Service Providers as reported by Spectrum Cellular.
[5] "On Data: Trying to Set the Standard," *Telocator*, July 1986.
[6] Robert Adair, Spectrum Information Technologies Executive Vice-President, from *Mobile Data Report*, 6-18-90.
[7] *En Route Technology*, 5-27-92.
[8] Frank Wapole, President and Chief Executive Officer of ARDIS, *Mobile Data Report*, 3-29-93.
[9] "[RAM will have] good in-building penetration in 20 . . . markets by the middle of 1993," Earl Mauldin, Executive Vice-President of Bell South Mobility, *Mobile Data Report*, 1-18-93.
[10] *En Route Technology*, 2-1-93.
[11] *Communications Week*, 2-24-92.
[12] Larry Sanders, RaCoTek Vice-President of Sales and Marketing, from *Mobile Data Report*, 2-10-92.
[13] Ibid.
[14] E*n Route Technology*, 1-19-93.
[15] Mike Fabiaschi, RaCoTek Vice-President of Sales/Support, from *En Route Technology*, 8-30-93.
[16] Larry Sanders, RaCoTek Vice-President of Marketing, from *En Route Technology*, 8-16-93.
[17] *Information Week*, 2-3-92.
[18] *Mobile Data Report*, 2-10-92.
[19] *Telecommunications Alert*, 1-23-92.
[20] *Mobile Data Report*, 2-24-92.
[21] *Industrial Communications*, 8-21-92.
[22] *Land Mobile Radio News*, 4-30-93.
[23] *Telecommunications Alert*, 5-4-93.
[24] *Telecommunications Alert*, 5-5-93.
[25] *Telecommunications Alert*, 6-10-93.
[26] *Communications Daily*, 9-1-93.
[27] Jeff Brown, McCaw Vice-President of Sales and Marketing, from *Communications Week*, 7-26-93.
[28] Rob Mechaley, Senior Vice-President and General Manager of McCaw Wireless Data Division, *Telocator Bulletin*, 4-30-93.
[29] *Mobile Data Report*, 6-21-93.
[30] *Telocator Bulletin*, 6-18-93.
[31] *Telocator Bulletin*, 7-2-93.
[32] *Mobile Data Report*, 6-21-93.

'DPD V0.8 Airlink Capacity Analysis,

eritech, 7-14-93.

CHAPTER 5

SUBSCRIBER GROWTH: HISTORY AND BARRIERS

5.1 APPROACH

Each of the public packet switched service providers reveals a different philosophy on reporting subscribers. The most successful, Omnitracs, is the most forthright about its user count; ARDIS, the next most successful, provides a reasonable—though sometimes inconsistent—track record. The least successful service providers, RAM and CoveragePLUS, are very guarded indeed.

Through 1991 a valuable source of information was the FCC loading records. Although this was only a register of *licenses*, not subscribers, it provided a useful sanity check. The FCC requirement to license new users has been dropped, and a valuable piece of marketing information has dropped with it. There is, however, some residual activity as existing systems are modified; this source has been spot checked for key customer names. The tables of names can be found in Appendices D–F.

The information gaps for 1992–1993 have been filled by news searches, examination of annual reports (where available), interviews with service providers, and other traditional techniques.

5.2 PUBLIC PACKET SWITCHED NETWORKS

5.2.1 ARDIS

ARDIS was formed with the merger of the Motorola public DRN network and the IBM private Field Service system (DCS). Its joint seven-year history is instructive when barriers to growth are examined.

The first task was to estimate the IBM/ROLM U.S. subscriber base through year-end 1992.

When DCS was created as a private system, IBM purchased 25,000 KDT800 terminals from Motorola. Only 21,000 of these units were actually put into service. Spare stocking, at the 5 percent level, put ~20,000 subscribers on the system in the peak year, 1985. That historical figure still appears in many trade journal articles.

Beginning in 1986, IBM U.S. employment began to decline.[1] It has been assumed that the Field Service Division followed this trend. The IBM U.S. employment and the estimated number of DCS subscribers, factored down at the same rate, are shown in Table 5-1.

Table 5-1 *ARDIS IBM Subscriber Count*

	1985	1986	1987	1988	1989	1990	1991	1992
U.S. employees	242,241	237,274	227,949	223,208	215,929	205,533	186,569	156,801
Est. service users	20,000	19,590	18,820	18,429	17,828	16,969	15,404	12,945

This estimate correlates well with responsible, externally reported ARDIS subscriber counts. The year-end 1989 value closely matches the January 1990 IBM statement.[2] The year-end 1990 value closely matches the 1991 "ARDIS Value Add" publication.[3]

The second task was to count the number of subscribers licensed on the three Motorola heritage (DRN) channel pairs in Chicago, Los Angeles, and Manhattan. These counts were quite useful in establishing early-year (pre-ARDIS) growth rates. Note that it is possible for a subscriber to be licensed but not yet active. Thus for some years the counts may be overstated, but they are directionally correct. The counts are recorded in Appendix D.

A summary of the detailed Motorola DRN heritage count is shown in Table 5-2.

Table 5-2 *Motorola DRN Subscriber Count*

	1985	1986	1987	1988	1989	1990	1991
Chicago	50	172	219	196	249	312	335
Los Angeles			180	214	331	658	661
Manhattan				37	191	511	589
Totals	50	172	399	447	771	1481	1585

The third step was to record subscriber unit quantities reported in the press or by ARDIS itself (see Table 5-3).

Table 5-3 *ARDIS Reported Subscribers*

	1991	1992	2Q93
NCR[a]	3,300	4,000[b]	4,000
Maersk	10	10	40[c]
ICL[d]	100	200	400
Pitney-Bowes[e]	3,200	3,200	3,200
AVIS[f]	157	157	175[g]
G.O.D.[h]	50	75	100
UPS	700[i]	1,500[j]	1,500
NYC Sheriff	10	10	100[k]
Otis[l]	—	2,000[m]	2,000
Sears[n]	—	125	500[o]
Unisys[p]	—	—	40
Great Western Bank	—	—	18[q]
Waste Management	—	—	17[r]
AT&T Paradyne	—	—	700[s]
Coca-Cola	—	68[t]	300[u]
Totals	7,527	11,345	13,090

[a] *En Route Technology*, 3-18-92.
[b] *On the Air*, Fall 1992, p. 3.
[c] *On the Air*, Spring 1993, p. 6.
[d] *Mobile Data Report*, 10-7-91.
[e] *En Route Technology*, 12-11-91.
[f] *Advanced Wireless Communication*, 9-18-91.
[g] *On the Air*, Winter 1993, p. 3.
[h] *Mobile Data Report*, 7-8-91.
[i] *Mobile Data Report*, 5-21-90.
[j] JFD discussion with Mary Jane Grinstead, ARDIS Senior VP, 9-2-92.
[k] *On The Air*, Winter 1993, p. 3; augmented at ARDIS Lexington Conference, 7-27-93.
[l] *En Route Technology*, 2-5-92.
[m] *On the Air*, Fall 1992, p. 5.
[n] *En Route Technology*, 12-26-91.
[o] *On the Air*, Spring 1993, p. 2.
[p] Atlanta, Baltimore/Washington tests, *En Route Technology*, 4-12-93.
[q] *On the Air*, Winter 1993, p. 2.
[r] Ibid, p. 6.
[s] Ibid, p. 7; augmented at ARDIS Lexington Conference, 7-27-93
[t] FCC loading record.
[u] ARDIS Lexington Conference, 7-27-93.

Note that there are 60 customers and 1,585 units recorded in 1991 on DRN heritage cities. Only 8 of the 60 customers (13 percent) were picked up via press reports. Approximately 20 percent of the known non-IBM devices were recorded in the top three cities, quite consistent with U.S. job distribution. Most of the residual 52 customers are city-specific, but several (Coca-Cola, Honeywell, Motorola, Tandem, UAL, Xerox) clearly have nationwide potential. At the end of 1992 only 10 of the 65 customers picked up on the FCC listings had reached press reports.

There are five other known but unenumerated users (ADP, Schindler, Shell Oil, Philadelphia Scofflaw, UARCO) that have been assigned to IBM heritage frequencies or to new frequencies added in key cities. This gives ARDIS a total of 70 customers at year-end 1992, an exact match with ARDIS's own press releases.[4]

Thus, current estimates of ARDIS users can be made with some confidence (see Table 5-4), suggesting that ARDIS is, to some degree, treading water as it continues to absorb the loss of IBM users.

Table 5-4 *ARDIS Subscriber Estimates: 1993*

	1992	2nd Quarter 1993
IBM Field Service	12,945	12,000
Press report tally	11,345	13,090
30% allowance for unreported	7,287	7,527
Totals	31,577	32,617

Is this possible? Consider the following ARDIS subscriber estimates:

1.	Active user projection (made 8/92) for 1/1/93[5]	40,000
2.	Users, fall 1992[6]	30,000
3.	Users, February 1993[7]	35,000
4.	Users, spring 1993[8]	32,000

This suggests goals not achieved, with some churning in the first quarter of 1993. Thus, the subscriber estimate made here appears reasonable.

With the formation of Bell-ARDIS, 1,100 IBM Canada Field Service personnel[9] and an assortment of mostly small Canadian test users can be added to the North American counts. The latter includes 300 Bell Canada users,[10] an undetermined number of subscribers from Otis, Ontario Hydro, and La Commission de la Construction du Quebec, as well as a solitary user from Hewlett-Packard. As IBM Canada is also scaling back, it would be generous to assume a Canadian total of 1,500 subscribers. (Note that Bell-ARDIS took responsibility for 75 Canadian base stations, which adds to the muddle of changing ARDIS base-station counts.)

Thus, the total ARDIS subscriber count represents ~4 percent utilization of the year-end 1991 channel capacity of 800,000 subscribers[11] on the four-channel/city bandwidth ARDIS has acquired.[12] With the addition of RD-LAP infrastructure, ARDIS utilization may be as low as 3 percent—which at least means they are well positioned to participate in the coming price wars.

A graph of subscriber history is shown in Figure 5-1.

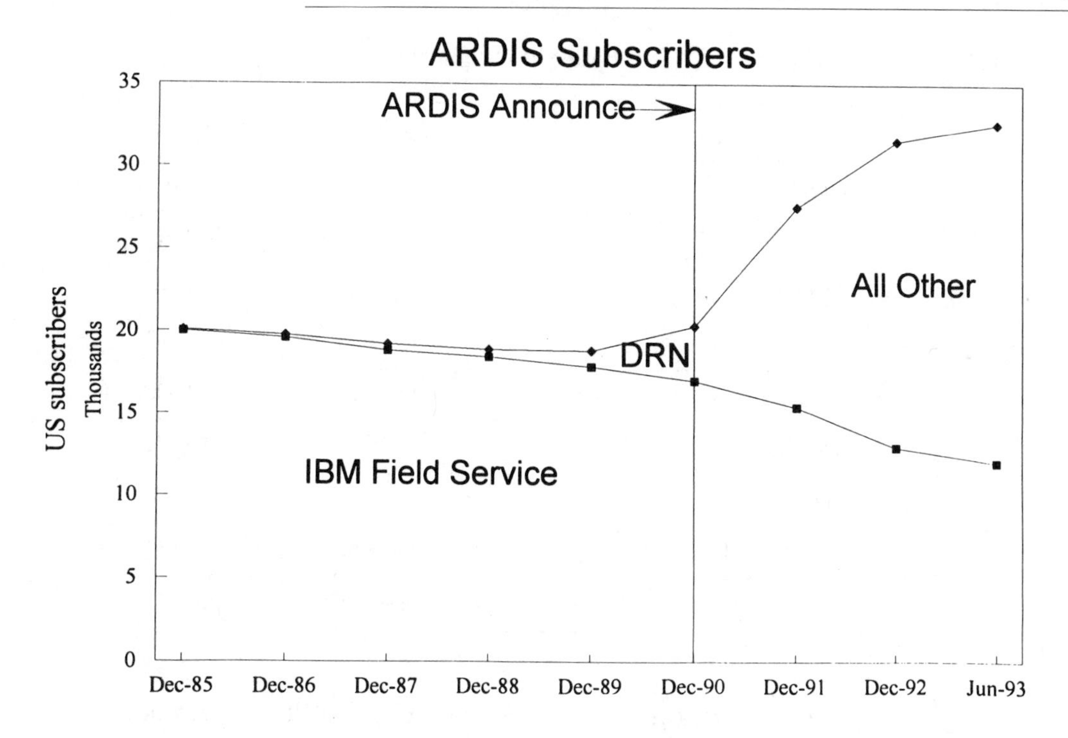

Figure 5-1 *ARDIS Subscribers: United States*

5.2.2 RAM

In October 1990 RAM became marginally operational with ten channels in each of ten metropolitan statistical areas (MSAs) across a total of 28 base stations.[13] No subscribers existed. RAM continues to be plagued by slow customer acceptance. City roll-out was initially slow. Reports of poor coverage were common[14] and persisted for at least two years. Competitive modems were unavailable.[15]

One coverage insight can obtained by noting the June 1993 New York City base station locations.[16] There are 9 "NYC" base stations compared to 12 for ARDIS. RAM locations are shown in Table 5-5.

Table 5-5 *RAM New York City Base Station Locations*

	Latitude	Longitude
Manhattan	40-45-25	73-58-12
Manhattan	40-42-28	74-00-16
Queens	40-43-06	73-48-47
Queens	40-46-06	73-52-07
Jamaica	40-39-29	73-46-17
Brooklyn	40-39-20	73-58-56
Brooklyn	40-39-52	73-57-32
Staten Island	40-35-57	74-06-55
Newark, NJ	40-45-08	74-11-10

Area comparisons are always imprecise, but it is worth noting that ARDIS has 145 base stations in the five boroughs, Westchester and Nassau counties, and New Jersey's Hudson shore.

RAM building penetration is currently not as rich as ARDIS, nor is this a surprise. In January BellSouth Mobility predicted that "RAM will have good in-building penetration in 20 of its 100 markets by the middle of 1993."[17] But its street-level coverage is likely quite competitive.

Another early inhibitor to RAM customer acceptance was lack of a competitive modem. Pacific Communication Sciences, Inc. (PCSI), had announced in early 1990 that it would have a product ready in 1991 in time for MSA expansion.[18] This did not happen. Further, Ericsson's Mobidem[19] was not yet commercially available; British Airways Speedwing prior airport test was conducted on three prerelease units.[20] The Mobidem problem has been more than solved; in its Intel labeled form it is currently the most competitively priced packet radio modem of its class.

As with ARDIS, an attempt was made to count licenses in the top three metro areas from FCC loading records. The results were unsatisfactory, complicated by the presence of SMR voice units on the RAM channels and fractional unit license balancing across multiple call signs and bases. The results for the period ending third quarter 1992 can be found in Appendix E.

Although there were no astonishing revelations, it was interesting to note the appearance of Anterior Technology (now RadioMail) as early as October 1992. About one-third of the licenses originate with Ericsson/GE or RAM itself, a phenomenon not unlike ARDIS with IBM and Motorola. GE Consumer Services does not yet appear in New York City or the other two major cities. A three-year rollout had been projected,[21] but the schedule may have been damaged by the financial collapse of Kustom and the subsequent purchase by Coded Communications.[22] It is also possible that early Mobidem shortages retarded rollout.

Press reports of active customers with quantities are very sketchy; samples are given in Table 5-6. Even this small sample had disappointments: American Courier Express's test results led to the decision to use alphanumeric pagers and voice SMR rather than mobile data terminals.[23]

Table 5-6 *RAM Press Reports: Quantifiable Subscribers*

	1991	1992	1st Half 1993	Potential
MB Trucking[a]	8			45
GE Consumer Services				2,500[b]–3,500[c]
Orlando County Sheriff[d]	20			850
Hoboken Fire Dep't.[e]	5			30
American Courier Express[f]		2		
Conrail			300[g]	
Bell Ambulance (Milwaukee)[h]			5	20
Regional Justice Info Service[i]				175

[a] *Advanced Wireless Communications*, 4-3-91.
[b] *Mobile Data Report*, 5-6-91.
[c] *Edge On and About AT&T*, 5-13-91.
[d] *Mobile Data Report*, 9-9-91.
[e] *Industrial Communications*, 10-11-91.
[f] *En Route Technology*, 2-19-92.
[g] RAM "Hard Data," Fall 1992, p. 2.
[h] *Communications Week*, 5-24-93.
[i] Ibid.

Press reports of unquantified users reveal only one high potential candidate: National Car Rental (Table 5-7) which announced it would place 1,500 to 2,000 employees (not vehicles) on RAM.[24]

Table 5-7 *RAM Press Reports: Unquantifiable Subscribers*

National Car Rental[a]
Hobart Field Service[b]
Chicago Parking Authority[c]
National Computer Systems[d]
Southern Cal Edison[e]
Zebra Air (Dallas)[f]
TransNet (Tampa)[g]

[55] *Advanced Wireless Communications*, 6-26-91.
[56] *En Route Technology*, 6-24-92.
[57] RAM "Hard Data," Fall 1992, p. 2.
[58] *Communications Week*, 5-24-93.
[59] Ibid.
[60] Ibid.
[61] Ibid.

These meager results led to the estimate of 1,000 users at year-end 1992 (YE92).[25] In fact, that judgment was optimistic. On March 29, 1993, the senior vice-president of RAM's Wireless Messaging Business Unit revealed that "we don't have many more than 1,000 (paying customers) today."[26]

However, RAM *must* not be taken lightly. Data radio is an excruciatingly hard business to build, and RAM is laying the most excellent foundations. Consider a few of its many business alliances:

1. MasterCard credit verification (1991)[27]
2. Hewlett-Packard (1992)
3. Microsoft Pen Windows (1992)
4. DEC All-in-1 E-mail (1992)[28]
5. AT&T Easylink E-mail (1992)[29]
6. Intel: 6,200 retail stores, potential for low-cost modems (1993)[30]

RAM airtime service is most aggressively priced. Its modems are the lowest priced and delivery constraints no longer exist. Metro-area coverage is good with 840 base stations in 210 major business centers.[31] It has a laser focus on promising horizontal applications such as electronic mail.

Thus, RAM may well have a fine 1993.

5.2.3 CoveragePLUS

Motorola began operations in "auto alley," the Chicago, Detroit, and Cincinnati triangle, in the fourth quarter of 1989. A year later CoveragePLUS was available on ~350 Motorola SMRs.[32] By second quarter 1991 more than 550 Motorola SMRs were linked,[33] permitting nationwide rollout to proceed with reasonable highway coverage. In the first quarter of 1993 all 750 Motorola SMRs were connected.[34]

Initial pilot testing was active: 160 customers were reported testing by fourth quarter 1990,[35] a number that contracted to 140 by second quarter 1991 (2Q91).[36] This was a result of both the start of operational rollout for some trucking firms, as well as a change of heart by others (e.g., Court Courier, originally announced for 70 trucks). These 140 firms have a 16,000-unit potential. Assuming a three-year rollout period, ~8,800 users could be operational by YE92. The predicted results are ~40 percent of that target.

The first step in estimating subscriber totals was to record all known users and quantities where responsibly reported. This search yielded the results in Table 5-8.

Table 5-8 *CoveragePLUS: Quantifiable Press Reports*

	YE90	YE91	YE92	1H93
Swift Transportation[a]	280	640	1,000	1,000
Roberson Corporation[b]	145	325	400	400
Harold Ives Trucking[c]	116	260	320	320
G&P Trucking[d]	—	—	—	225
Hi-Way Dispatch[e]	35	95	155	155
D&J Transfer[f]	58	100	100	100
B. R. Williams[g]	—	25	49	77
Alton Bean[h]	—	—	64	64
Mobile Billboards[i]	—	50	50	50
R. Wayne Bost[j]	11	32	32	32
C. H. Dredge[k]	—	—	—	110
Totals	645	1,527	2,170	2,533

[a] Motorola 4/89 CoveragePLUS announcement: "Swift Transportation of Phoenix, AZ has signed on for 1,000 units." The rollout schedule has not been publicly printed; full rollout by YE92 was assumed.
[b] *Industrial Communications*, 5-1-92.
[c] Ibid.
[d] *Communications Daily*, 8-27-92.
[e] *Mobile Data Report*, 4-9-90. Fifteen units were in test with a full potential of 155 units; rollout commenced in the fall of 1990 and was judged complete by YE92.
[f] *Mobile Data Report*, 7-1-91.
[g] Ibid.
[h] *En Route Technology*, 2-5-92.
[i] *Mobile Data Report*, 7-1-91.
[j] Ibid.
[k] *En Route Technology*, 8-2-93.

But the public record is lean. The customer assumptions shown in Table 5-9 were made.

Table 5-9 *CoveragePLUS: Pilot versus Operational Assumptions*

	1990	1991	1992	1H93
Pilot	150	126	103	80
Known live	6	8	9	10
Unknown live	4	6	8	10
Totals	160	140	120	100
Cumulative dropped	—	20	40	60

The number of units per subscriber was estimated as follows:

1. Pilot = 3 units
2. Known user = Table 5-8
3. Unknown user = 50 units

That yields a subscriber estimate of the magnitude shown in Table 5-10.

Table 5-10 *CoveragePLUS: Subscriber Estimates*

	1990	1991	1992	1H93
Pilot	450	378	309	240
Known user	645	1,527	2,170	2,533
Unknown user	—	300	400	500
Totals	1,095	2,205	2,879	3,273

This estimate suggests that CoveragePLUS is making very slow progress (see Figure 5-2). However, it is in line with independent estimates made by Qualcomm.[37] But CoveragePLUS could receive a boost from minority-held RaCoTek with its new

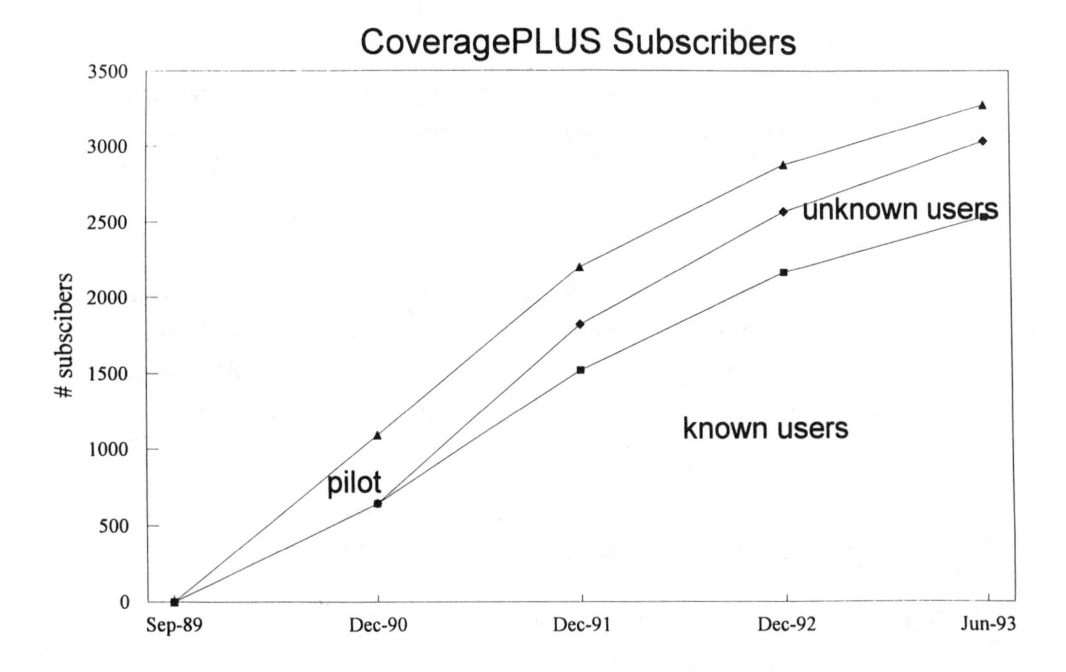

Figure 5-2 *CoveragePLUS Subscribers*

offering: RaCoNetPLUS.[38] This hardware and software development effort integrates RaCoNet with CoveragePLUS. With the addition of American Freightways on RaCoNet,[39] about 2,000 additional users can be counted on the combined networks.

CoveragePLUS revenue is also probably low. Many optional services are not being purchased. The minimum monthly charge is $35 and the national average monthly bill is only $40–45.[40]

It must be pointed out, however, that CoveragePLUS now covers 90 percent of the interstates with a response time less than 10 seconds.[41] Motorola currently has enough faith in the program to convert at least 500 transmitter sites to digital.[42] Stay tuned.

5.2.4 Omnitracs

Though it is not a terrestrial system, it would be a serious omission to ignore Qualcomm's satellite-based Omnitracs system. A representative list of users can be found in Appendix F. This is not meant to be exhaustive but simply a record of Omnitracs orders/installs captured while searching for terrestrial information. But it tallies ~80 percent of the reported U.S. units. A cursory glance at this list reveals also the emphasis on IBM hosts (3090s, A/S 400s, R/S 6000s) and even IBM terminal devices: the OnBoard Computer (OBC), now called the "Roadrunner."[43]

A gross summary, mostly from quarterly financial reports, of the U.S. installed position, is shown in Table 5-11.

Table 5-11 *Omnitracs U.S. Installed Position*

Installed	Quantity	Source	Date
10/89	6,000	*Industrial Communications*	11-10-89
5/90	7,500	*Mobile Data Report*	6-18-90
9/90	8,600	*Mobile Satellite News*	9-90
1/91	12,000	*Inside IVHS*	2-4-91
3/91	13,500	*Mobile Data Report*	3-11-91
1/92	22,000	*En Route Technology*	2-5-92
7/92	29,000	*En Route Technology*	8-5-92
3/93	35,000	*Communications Daily*	4-7-93
6/93	45,000	*Mobile Satellite Reports*	7-5-93

The growth curve is depicted in Figure 5-3. Note that this graph does not include units shipped outside the U.S. The 45,000 units almost certainly include maintenance spares and the number is perhaps 5 percent richer than actual subscribers. But it does not include IBM units testing on Omnitracs (85 at J. B. Hunt alone)[44] so it is directionally correct.

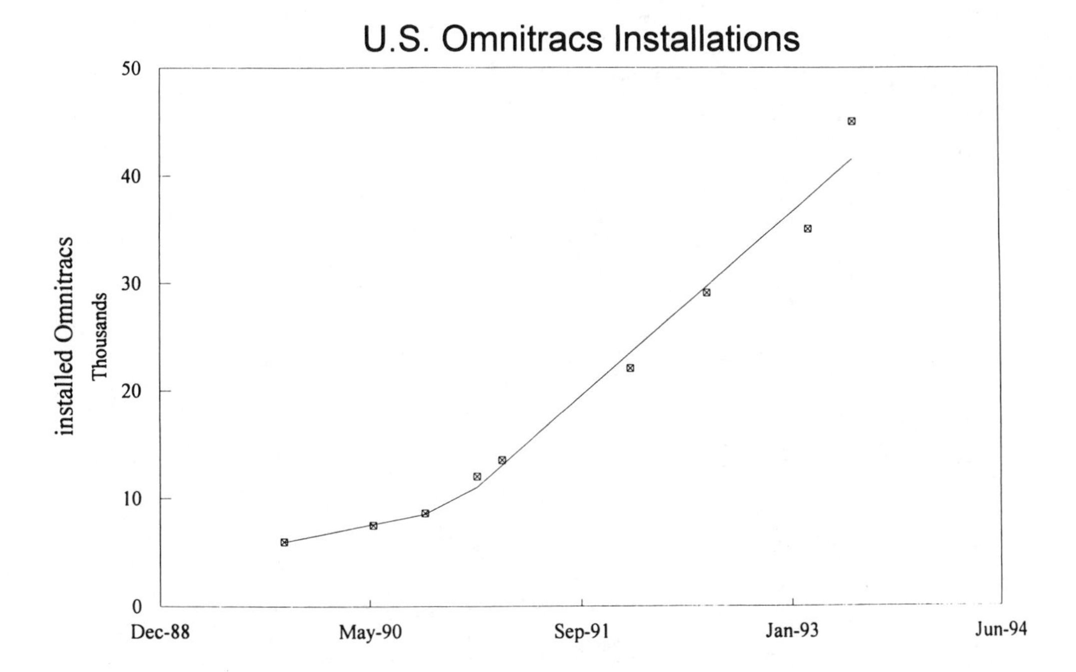

Figure 5-3 *U.S. Omnitracs Installed*

Omnitracs achieved more than 200 customers by the second quarter of 1993,[45] compared to 70 for ARDIS. Its subscriber base is also larger than ARDIS. But it is useful to note that the base has built at a much slower pace than originally projected. The announcement outlook was for 30,000 units by YE89 and 50,000 units by YE90.[46] Data radio is a tough business.

Qualcomm lost money throughout 1991 and 1992. But the Omnitracs division is reported to have been profitable for more than a year[47] (parent losses came from the CDMA business). In 1993 Qualcomm became profitable, due in part to CDMA contracts but also "to a sharp rise in the sale of its Omnitracs . . . terminals."[48] The Omnitracs installed base is now large enough that the airtime service revenues approach those of hardware—an important consideration if IBM's far more functional devices begin to trim the sale of Omnitracs units.

This hardware/airtime revenue equivalency can be coarsely estimated from the financial figures shown in Table 5-12.

			Fiscal 1991				Fiscal 1992				Fiscal 1993		
	3Q	4Q	1Q	2Q	3Q	4Q	1Q	2Q	3Q	4Q	1Q	2Q	3Q
Quarter ending	6/90	9/90	12/90	3/91	6/91	9/91	12/91	3/92	6/92	9/92	12/92	3/93	6/93
Sales ($M)	8.5		17.9	21.0	22.2		25.2	26.3	27.1	29.0	30.0	36.8	49.1
P/L ($M)			(.24)	(.55)	(.60)		(1.7)	(1.2)	(.90)	(.60)	(1.6)	.4	6.1
Quarter ship ww			1,589				3,551	3,837			4,865	5,155	
Annual ship ww			8,831				14,879						
Cum. ship (ww)	8,600						22,000		29,000		38,000	43,000	

Table 5-12 *Qualcomm Financial History*

Thus,

~5200 units/quarter × $4,000/unit = $20,800,000
hardware revenue/quarter

43,000 units × $450 airtime/quarter = $19,350,000
airtime revenue/quarter

5.3 DATA-OVER-CELLULAR

No precise count of data users on cellular is possible because, with some exceptions, the call is handled as if it were voice.

The range of modem types known to be using cellular is as wide as anything encountered on wireline systems. Parsons Bromfield-Redniss & Mead, a surveying firm in Fairfield County, Connecticut, use acoustic couplers with very low speed modems to successfully transmit data between laptop and host, via cellular phones. The ubiquitous Microcom networking protocol (MNP) modems also work well with cellular, and at consistently higher transmission speeds.

The same generalities hold true for facsimile transmission via cellular. If the speed is held down to, say, 4800 bps, facsimile works well with fax modems and cellular phones with a data attachment capability—and the call cannot be distinguished from voice.

Estimates of the number of cellular data users are wide ranging, but a firm floor has been established with UPS. In the first quarter of 1993, ~50,000 trucks were equipped with dedicated data devices,[49] more than either ARDIS or Qualcomm Omnitracs. Recent reports[50] indicate the UPS total may have increased another 10 percent. Thus, ~.5 percent of all cellular subscribers are UPS data users, and more if one focuses on only the business user component of cellular subscribers.

Recent consultants' estimates have been edging upward. One example is the conservative view of Herschel Shosteck Associates[51]: "about 2 percent of the cellular subscribers . . . are transmitting data *including facsimile* [italics mine]."

Carriers have also begun to report estimates of cellular data or facsimile usage. Examples:

1. Sprint: 3–10 percent of *subscribers.*[52]
2. Bell Atlantic Mobile: 3 percent of *traffic* (far more meaningful).[53]
3. Bell Mobility (Canada) projects that "2 to 3 percent . . . of [their] customers may be using the system for data transmission by the end of (1993)."[54] Further, "within 5 years as much as 20 percent of the . . . *traffic* could be data."

Further, the carriers have begun to act on their estimates:

1. Sprint has created a national cellular data team with 40 managers and 160 salespeople to "develop existing markets rather than concentrating on new technologies."[55]
2. Bell Atlantic Mobile has announced[56] that it expects "data traffic will increase another 2 to 4 percent within a couple of years" and expects data to contribute "more than 5 percent of total revenues within the next 12 to 24 months." Some of this traffic is to be captured by "greater marketing and better services for circuit switched cellular." BAMS intends to "offer new services and prices for ciruit switched data" including "modem pools . . . and pricing in less than one minute increments."
3. Bell Mobility "will offer rates based on fractions of a minute." It is also cooperating with Bell-ARDIS (60 percent owned by BCE Mobile): "in the future the two companies might work together to sell data services."

But optimistic carrier views have been seen before. It is true that a positive outlook is all the more urgent in the face of declining average monthly *voice* billings. However, it is useful to reflect on the carriers' history with data over cellular:

1. In the spring of 1985 Bell Atlantic began piloting with 20 organizations, using 3 different laptop models and the CTS1620 modem.[57]
2. In April 1986 "customer surveys conducted by the cellular subsidiaries of several Bell operating companies show that 5 to 17 percent of current voice subscribers would use cellular data transmission facilities if they were available."[58]
3. In March 1987 Ameritech offered the first "311" U.S. data service.[59]
4. In 1988 Bell South Mobility stated: "the 90's will definitely be the decade of data . . . More than 15 percent of our customers have indicated a strong interest in data communications service."[60]
5. In 1989 Cincinnati Bell signed an agreement for exclusive marketing rights to Agilis in the public safety application area where "the market is virtually untapped."[61]

Nor were the carriers alone in their hopes:

1. CompuServe had five operational "cellular cities" by February 1986.[62]
2. In the spring of 1989 the Internal Revenue Service (IRS), which had already made a major laptop purchase, began to explore data via cellular.[63]
3. Six months later the Los Angeles Police Department (LAPD) announced plans to test 25±5 laptops via cellular commencing 1Q90. The market potential was estimated to be 3500±500 units in an application already using ~850 Electrocom Automation devices.[64]
4. NovAtel stated in September 1989, "a large percentage of data transmissions within the next several years will be cellular."[65]
5. In October 1990, EDS bought Appex to catch the new opportunity.[66]
6. Nearly simultaneously the California Trucking Association announced a plan for 2,500 members to use data-over-cellular.[67]

7. In 1991 Information Systems, Inc., introduced its Computer Automated Transmission system, based squarely on Tandy products: PCs, integrated modem, and cellular phone, tied together with Tandy's data adapter, which also could handle a Ricoh fax.[68]

What is the impact of eight years of work? Here are some clues:

1. By April 1986, after more than a year of intense effort, Spectrum Cellular had only been able to sell 1,000 Bridge/Span units;[69] by June 1990, more than five years after first customer ship the Spectrum cumulative total had reached only 12,000 units.[70] In 1991 GTE abandoned testing of Spectrum's Bridge and Span because it "wasn't sufficiently impressed . . . to include [them] at the MTSO. If a customer wants to transmit data over cellular we won't stand in their way, but it's not cost effective."[71]
2. In February 1990, Cellular Solutions' carefully targeted Multiple Listing Service system for real estate agents failed in spite of the choice of the agent-familiar T.I. 707 terminal and the fact that one-third of the agents already had cellular phones.[72] Reasons given: "resistance, high turnover."
3. In 1992 Shell Oil ceased work on its Houston cellular data network because data "was getting splattered by noise."[73]
4. Eastman Kodak, after a four-city "technical success" using MNP10 modems to save a half hour per day of a service technician's time, did not proceed with the project because they could not offset the $250/month cost by time saved.[74]

Interestingly, some of the most conservative views have come from hardware vendors:

1. Motorola, spring 1985: "this market is minuscule and isn't likely to grow for a long time."[75]
2. AT&T Safari, March 1992: "marketing of [wireless] options is going slow, very slow . . . and it doesn't look like it will pick up any time soon."[76]

3. DEC, June 1993: "it will take three to five years before wireless computing enters the mainstream."[77]

Obviously, there have been success stories. However, general-purpose users tend to be for small quantities (e.g., Berrien County, MI Department of Social Services, four integrated modem, 2400 bps Zenith SuperSports via Datajack and Motorola/Uniden phones, each used 3–4 times/week;[78] Jersey Central Power & Light work orders),[79] usually where the need is application specific (e.g., ABS Courier Services, Fairfax, VA).[80]

The core problem is no longer transmission coverage, reliability, product cost, or "too many boxes." Quite simply, the MNP or V.42 modem that works with cellular works even better on landlines—and landlines are everywhere. The mobile need is quite often well served by the ubiquitous RJ-11C in the wall.

Today's cellular data[81] and fax subscribers sharply outnumber their packet switched equivalents, probably by at least 3 to 1. Further, casual cellular data/fax users are likely to grow at a far faster pace than packet switched. But a once-a-week cellular facsimile user is simply not a major revenue producer. Cellular data is shaping up to be a case of many, many low-intensity subscribers—making comparisons to packet switched more difficult.

REFERENCES

[1] IBM Annual Reports.
[2] IBM, January 30, 1990, press release: "The [DCS] network is now used by 16,000 IBM and 2,000 ROLM service personnel."
[3] ARDIS sales information accompanying the June 1991 *Datapro Report*: "ARDIS . . . provides communication between more than 15,000 IBM field engineers."
[4] Open letter to Mobile '93 attendees from Frank Wapole, President and Chief Executive Officer of Ardis, 2-26-93.
[5] ARDIS "Quick Reference," presented to JFD Associates, 10-21-92.
[6] Frank Wapole quoted in *On the Air*, Fall 1992, p. 2.

[7] Frank Wapole letter to Mobile '93 attendees, 2-26-93.
[8] "The Power of ARDIS," *On the Air*, p. 2.
[9] *En Route Technology*, 2-19-92.
[10] *En Route Technology*, 4-26-93.
[11] ARDIS press release as reported in *Mobile Data Report*, 12-2-91.
[12] Jack Blumenstein, ARDIS President, as quoted in *Mobile Data Report*, 1-27-92.
[13] *Mobile Data Report*, 10-8-90.
[14] *Mobile Data Report*, 2-10-92.
[15] *Telecommunications Alert*, 1-30-92.
[16] FCC loading records.
[17] Earl Mauldin, Group President of Mobile Systems, Bell South Enterprises, *Mobile Data Report*, 1-18-93.
[18] Michael Lubin, PCSI Executive Vice-President, *Mobile Data Report*, 4-23-90.
[19] *Mobile Data Report*, 2-10-92.
[20] *Information Week*, 2-3-92.
[21] Charles Bowen, Kustom Marketing Vice-President, *Mobile Data Report*, 1-15-90.
[22] *En Route Technology*, 11-27-91.
[23] *En Route Technology*, 8-5-92.
[24] *Advanced Wireless Communications*, 6-26-91.
[25] JFD Associates, *Wireless Data Handbook*, 2H92 edition.
[26] Martin Levetin, *Mobile Data Report*, 3-29-93.
[27] *Mobile Data Report*, 8-12-91.
[28] *Advanced Wireless Communications*, 6-10-92.
[29] *Electronic Messaging News*, 7-8-92.
[30] *Common Carrier Week*, 2-22-93.
[31] *Telecommunications Alert*, 6-23-92.
[32] *Mobile Data Report*, 11-19-90.
[33] *Mobile Data Report*, 7-1-91.
[34] *En Route Technology*, 3-15-93.
[35] Pat Schod, Motorola, as reported in *Mobile Data Report*, 11-19-90.
[36] Pat Schod, Motorola, as reported in *Mobile Data Report*, 7-1-91.
[37] Tom Bernard, Senior Vice-President and General Manager, Qualcomm, *En Route Technology*, 3-15-93.
[38] *Communications Alert*, 8-10-93.
[39] *En Route Technology*, 8-30-93.
[40] Jeff Templeton, National Sales Manager, CoveragePLUS, *En Route Technology*, 3-15-93.
[41] *En Route Technology*, 3-15-93.
[42] *Land Mobile Radio News*, 2-26-93.

[43] *En Route Technology*, 2-1-93.

[44] Ibid.

[45] *Satellite News*, 4-5-93.

[46] Allen Salmasi, Omninet President and Founder, reported in *Satellite News*, 1-25-88.

[47] *En Route Technology*, 8-5-92; *Satellite News*, 12-21-92.

[48] *Communications Alert*, 7-20-93.

[49] Paul Heller, UPS Division Manager for Mobile Networks, from *Mobile Data Report*, 6-7-93.

[50] *En Route Technology*, 4-12-93.

[51] *Mobile Data Report*, 12-21-92.

[52] *Mobile Data Report*, 6-7-93.

[53] *Mobile Data Report*, 6-1-93.

[54] David Roth, Project Manager for Data Services, *Mobile Data Report*, 12-21-92.

[55] *Mobile Data Report*, 6-7-93.

[56] Benjamin L. Scott, Executive Vice-President and Chief Executive Officer of BAMS, *Mobile Data Report*, 6-21-93.

[57] *Information Week*, 5-6-85.

[58] LOTUS, 4-86.

[59] *Radio Communications Review*, 3-15-87.

[60] Betty Rhodes, Manager, Advanced Cellular Products, Bell South Mobility, as reported in *Cellular Marketing*, Spring 1991.

[61] Ron Doepker, Cincinnati Bell Systems Manager, in *Mobile Data Report*, 8-14-89.

[62] *Mobile Phone News*, 2-12-86.

[63] *Mobile Data Report*, 3-15-89.

[64] *Mobile Data Report*, 9-25-89.

[65] *Mobile Product News*, 9-89.

[66] *Communication Systems News*, 10-8-90.

[67] *Mobile Data Report*, 10-8-90.

[68] *Mobile Data Report*, 2-11-91.

[69] *Linkup*, 4-86.

[70] Bob Adair, Executive Vice-President of Spectrum Information Technologies, *Mobile Data Report*, 6-18-90.

[71] Randy Crouse, GTE Director of Technology Planning, *Mobile Data Report*, 1-14-91.

[72] *Mobile Data Report*, 2-26-90.

[73] Joe Marsh, Shell Oil Manager of Communications Technology, *Advanced Wireless Communications*, 6-9-93.

[74] Bob Derounian, Kodak Project Manager, reported in *Advanced Wireless Communications*, 6-9-93.

[75] James Caile, Motorola Strategic Marketing Manager, reported in *Information Week*, 5-6-85.

[76] John Cowley, Mobile Communications Manager, Safari Systems, *Mobile Data Report*, 3-9-92.

[77] Jim Slane, Business Manager of DEC Mobile & Wireless Group, *Mobile Data Report*, 6-21-93.

[78] *Mobile Data Report*, 3-11-91.

[79] *Information Week*, 10-19-92.

[80] *Mobile Product News*, 5-25-89.

[81] BIS Strategic Decisions estimated 25,000 data users in 1991 as reported in *Edge On & About AT&T*, 2-3-92.

CHAPTER 6

MARKET OPPORTUNITY

6.1 THE SECOND ERA OF LOW-HANGING FRUIT

Many businesspeople are captivated by enormous market opportunity estimates for data. AT&T "projects 1 billion people will use wireless modems for electronic data communications by the year 2000."[1] That's one-third the population of the earth, only seven years hence. We'd all better get busy.

Some recent estimates are more plausible, though far from agreement (~4:1 spread):

Source		Date	Prediction	
IDC[2]	2/93	1997	5,000,000	Wireless WAN users
EMCI[3]	2/93	1997	9,600,000	2-way mobile data users
BIS[4]	6/93	1998	3,926,000	Mobile data users
Link Resources[5]	6/93	1998	4,208,000	PICs
Ovum, Ltd.[6]	6/93	1998	2,670,000	40% packet, 60% cellular
Yankee Group[7]	6/93	1998	7,300,000	Mobile data, mobile office, PDAs

Spurred by these tantalizing quantity projections, multiple service providers are scrambling to bring enormous capacity online. A reality check is in order.

6.2 SOME UNPLEASANT HISTORY

In 1987 a unique study of the U.S. data radio opportunity was completed.[8] This report had an unusually large circulation, but its conclusions on the finiteness of the data radio market opportunity were not lovingly received.

The approach was to build a detailed database of occupations, with principal source information obtained from the Bureau of Labor Statistics Occupation Industry matrix. This information was supplemented by census data to make self-employment estimates. The 579 detailed occupation codes were initially compressed to 408 categories; 3 sensitive occupation codes were expanded to 26 from other data sources. The result was 434 detailed occupation codes organized for bottom-up analysis. Two levels of summary information were provided for ease of use. Projections to 1991 and 1996 were made from the 1986 base.

This work was aimed at public data service providers, not hardware vendors. The object was to estimate the Public Data Wide Area Mobile (WAM) opportunity for business users: that is, data radio users who ranged over a wide area, as opposed to in-building or campus, and who would use public services, as opposed to private. The focus was on business jobs because, except for the deaf, few nonbusiness data radio users were anticipated. The approach used in a subsequent JFD Associates update was as follows:

1. All 434 occupations were first evaluated to determine the *mobile* content. Thus, barbers, podiatrists, dental hygienists, and so on were rejected as mobile candidates.
2. Mobile jobs were examined for classification as *campus/in-building* or *wide area.* Thus, stock clerks, waiters, lobby attendants, and so on were mobile, but not in a wide area. An obvious contrast was long haul tractor-trailer drivers who are clearly not campus-confined. These detailed judgments were not binary. High school principals were 95 percent campus, but there were exceptional cases where the principal moves from school to school. There were no double counts. A worker mobile in both categories was classified as WAM.
3. Judgments were made as to which jobs use *two-way radio,* either voice or data. Fishing vessel captains, police, and firefighters were naturals; very few actors were expected to use radio in their work.
4. The two-way radio users were classified for *public* versus private. Health services managers were very likely to choose public systems, if available. But subway operators or fire inspectors will tend to be on private systems.
5. Finally, the wide area mobile, two-way, public radio users were examined to determine who fits *data* as opposed to voice (one can have both). Electronic repair personnel were high data; motorcycle mechanics have low data utility.

The result was a 125-page report with 6 statistical appendices that quantified the opportunity in both 1991 and 1996. The 1991 job estimate, based upon 1986 data, was slightly pessimistic, but still accurate. The predicted job count versus the actuals reported by the Bureau of Labor Statistics:[9]

BLS actual: 1/89	116,708,000
JFD predicted: 12/91	115,927,360
BLS actual: 8/92	117,737,000

6.2.1 Summary Examples: 1991–1996

An illustrative example of the detailed job counts is in order.

Table 6-1 is a single-page extract from the third appendix, exploded for clarity. It contains an estimate that the total number of mobile workers in 1991 would be 23,340,856 + 24,825,878 = 48,166,734. The key point was the separation of wide-area mobile workers, who are the fundamental target for any public airtime service, from in-building/campus, who represent the Altair/WaveLAN opportunity (the WAM segment is just under half the mobile job total).

This crucial distinction is often ignored when the report is quoted with unhelpful statements such as "there are 48 million mobile workers" in the United States.[10]

Many other insights can be extracted from just this single page: 100 percent of both school principals and surveyors were mobile. But 95 percent of the principals were campus, of little value to a public carrier, while 90 percent of the surveyors were wide area, which is pertinent. Also obvious are the large job totals associated with management personnel. This could be fertile ground for wireless communication given the vast improvement in laptop computers—but it is also an area suffering from heavy job attrition as the economy goes through the wringer.

					Wide Area Mobile				In-bldg, Campus Mobile		
BLS CODE	BLS SEQ	SEG	Description	% WAM	Number in 1986	Number in 1991	Number in 1996	% Campus	Number in 1986	Number in 1991	Number in 1996
				20.2%	21579337	23340856	24983772	21.3%	22804242	24825878	26736858
15003	4	11	Assistant principals	5	1990	2097	2190	95	37819	39840	41605
15004	5	11	Principals	5	4335	4567	4769	95	82370	86772	90619
15026	6	11	Managers, food service and lodging	15	85012	94188	102588	85	481734	533734	581331
15002	7	11	Postmasters and mail superintendents	10	2810	2870	2920	90	25288	25826	26282
19002	8	11	Public administration chief executives, legislative, and general administration	20	28482	30294	31892	80	113928	121177	127569 19999
13014	10	11	Managers, administrative services	10	21914	24946	27792	30	65741	74839	83376
15023	11	11	Managers, communications, transportation and utilities operations	25	43063	49023	54615	30	51676	58828	65538
15017	12	11	Managers, construction	50	96384	109724	122239	30	57831	65834	73344
15005	13	11	Administrators, education	15	42064	47886	53348	60	168256	191543	213391

Table 6-1 *Mobile Job Counts for Detail Occupations (sample page Appendix C)*

					Wide Area Mobile				In-bldg, Campus Mobile		
BLS CODE	BLS SEQ	SEG	Description	% WAM	Number in 1986	Number in 1991	Number in 1996	% Campus	Number in 1986	Number in 1991	Number in 1996
13017	14	11	Managers, engineering, mathematical, and natural sciences	15	7582	8632	9616	60	30330	34527	38466
13002	15	11	Managers, financial	10	65460	74519	83019	30	196379	223557	249058
19005	16	11	General managers and top executives	25	638639	727025	809955	50	1277278	1454050	1619909
15014	17	11	Managers, industrial production	10	14207	16173	18018	90	127864	145559	162163
13011	18	11	Managers, marketing, advertising, and public relations	15	50610	57614	64186	30	101219	115228	128371
15008	19	11	Managers, medicine and health services	15	16065	18289	20375	85	91036	103635	115457
15021	20	11	Managers, mining, quarrying, and oil and gas well drilling	15	7582	8632	9616	85	42967	48914	54493
13005	21	11	Managers, personnel, training, and labor relations	10	15997	18211	20288	30	47991	54633	60865
15011	22	11	Managers and administrators, property and real estate	50	25275	28773	32055	30	15165	17264	19233

Table 6-1 *Mobile Job Counts for Detail Occupations (sample page Appendix C) (continued)*

					Wide Area Mobile				In-bldg, Campus Mobile		
BLS CODE	BLS SEQ	SEG	Description	% WAM	Number in 1986	Number in 1991	Number in 1996	% Campus	Number in 1986	Number in 1991	Number in 1996
13008	23	11	Managers, purchasing	15	36779	41869	46645	30	73559	83739	93291
21902	30	12	Cost estimators	10	11738	12880	13916	10	11738	12880	13916
22102	52	12	Aeronautical and astronautical engineers	10	4973	5727	6442	10	4973	5727	6442
22114	53	12	Chemical engineers	15	8747	9831	10836	10	5831	6554	7224
22121	54	12	Civil engineers, including traffic engineers	25	44965	51127	56901	10	17986	20451	22760
22126	55	12	Electrical and electronics engineers	10	40485	51024	61874	10	40485	51024	61874
22128	56	12	Industrial engineers, except safety engineers	10	12753	14694	16537	10	12753	14694	16537
22135	57	12	Mechanical engineers	10	24318	28550	32635	10	24318	28550	32635
22105	58	12	Metallurgists and metal, ceramic, and material engineers	10	1983	2218	2435	10	1983	2218	2435
22108	59	12	Mining engineers, including mine safety engineers	10	634	656	675	10	634	656	675

Table 6-1 *Mobile Job Counts for Detail Occupations (sample page Appendix C) (continued)*

					Wide Area Mobile				In-bldg, Campus Mobile		
BLS CODE	BLS SEQ	SEG	Description	% WAM	Number in 1986	Number in 1991	Number in 1996	% Campus	Number in 1986	Number in 1991	Number in 1996
22117	60	12	Nuclear engineers	10	986	1032	1071	10	986	1032	1071
22111	61	12	Petroleum engineers	10	2285	2488	2671	10	2285	2488	2671
22132	63	12	Safety engineers, except mining engineers	10	1303	1498	1684	10	1303	1498	1684
22199	64	12	All other engineers	10	22896	26484	29900	10	22896	26484	29900
22301	69	12	Architects, including landscape architects	10	9968	11362	12671	10	9968	11362	12671
22309	72	12	Surveyors	90	41519	44412	46977	10	4613	4935	5220
25102	76	12	Computer systems analysts, electronicnic data processing	10	31458	41960	53345	10	31458	41960	53345

Table 6-1 *Mobile Job Counts for Detail Occupations (sample page Appendix C) (continued)*

Table 6-2 is a single-page extract from the sixth appendix, exploded for clarity. WAM jobs alone are not sufficient for a wireless data business evaluation. Several other screens act as a filter on the potential. Does the job require a two-way radio (any kind: SMR voice dispatch through private packet); will that two-way radio job be performed on a public network; is the public job a candidate for data (quite possibly with voice). Thus, in 1991 surveyors shrank from 44,412 to 14,345 when these criteria were applied. And this is an application area that *is* modernizing; coordinate input and facsimile output via data over cellular are happening today.

					1986		1991		1996	
BLS CODE	**BLS SEQ**	**SEG**	**Description**	**% Data**	**# Public**	**# Data**	**# Public**	**# Data**	**# Public**	**# Data**
			Totals:	20.0%	5927406	1188381	6494338	1317046	7022814	1437732
15003	4	11	Assistant principals	20%	945	189	996	199	1040	208
15004	5	11	Principals	20%	2059	412	2169	434	2265	453
15026	6	11	Managers, food service and lodging	25%	40381	10095	44739	11185	48729	12182
15002	7	11	Postmasters and mail superintendents	20%	843	169	861	172	876	175
19002	8	11	Public administration chief executives, legislative, and general administration	20%	10823	2165	11512	2302	12119	2424
19999	9	11	All other managers and administrators	25%	121563	30391	138387	34597	154173	38543
13014	10	11	Managers, administrative services	25%	10409	2602	11849	2962	13201	3300
15023	11	11	Managers, communications, transportation and utilities operations	35%	21532	7536	24512	8579	27308	9558
15017	12	11	Managers, construction	25%	31325	7831	35660	8915	39728	9932

Table 6-2 *Wide Area, Public, Data Radio Opportunity (sample page Appendix F)*

					1986		1991		1996	
BLS CODE	BLS SEQ	SEG	Description	% Data	# Public	# Data	# Public	# Data	# Public	# Data
15005	13	11	Administrators, education	20%	15984	3197	18197	3639	20272	4054
13017	14	11	Managers, engineering, mathematical, and natural sciences	20%	3602	720	4100	820	4568	914
13002	15	11	Managers, financial	25%	21765	5441	24778	6194	27604	6901
19005	16	11	General managers and top executives	25%	303354	75838	345337	86334	384728	96182
15014	17	11	Managers, industrial production	25%	12147	3037	13828	3457	15405	3851
13011	18	11	Managers, marketing, advertising, and public relations	25%	21636	5409	24630	6157	27439	6860
15008	19	11	Managers, medicine and health services	30%	11446	3434	13031	3909	14517	4355
15021	20	11	Managers, mining, quarrying, and oil and gas well drilling	25%	1896	474	2158	539	2404	601
13005	21	11	Managers, personnel, training, and labor relations	20%	3799	760	4325	865	4818	964
15011	22	11	Managers and administrators, property and real estate	30%	14407	4322	16401	4920	18271	5481

Table 6-2 *Wide Area, Public, Data Radio Opportunity (sample page Appendix F) (continued)*

					1986		1991		1996	
BLS CODE	BLS SEQ	SEG	Description	% Data	# Public	# Data	# Public	# Data	# Public	# Data
13008	23	11	Managers, purchasing	40%	15723	6289	17899	7160	19941	7976
21902	30	12	Cost estimators	30%	5576	1673	6118	1835	6610	1983
22102	52	12	Aeronautical and astronautical engineers	20%	1654	331	1904	381	2142	428
22114	53	12	Chemical engineers	20%	4986	997	5604	1121	6177	1235
22121	54	12	Civil engineers, including traffic engineers	25%	26979	6745	30676	7669	34141	8535
22126	55	12	Electrical and electronics engineers	25%	19231	4808	24236	6059	29390	7347
22128	56	12	Industrial engineers, except safety engineers	25%	6057	1514	6980	1745	7855	1964
22135	57	12	Mechanical engineers	25%	8086	2021	9493	2373	10851	2713
22105	58	12	Metallurgists and metal, ceramic, and material engineers	25%	754	188	843	211	925	231
22108	59	12	Mining engineers, including mine safety engineers	25%	228	57	236	59	243	61
22117	60	12	Nuclear engineers	25%	375	94	392	98	407	102

Table 6-2 *Wide Area, Public, Data Radio Opportunity (sample page Appendix F) (continued)*

					1986		1991		1996	
BLS CODE	BLS SEQ	SEG	Description	% Data	# Public	# Data	# Public	# Data	# Public	# Data
22111	61	12	Petroleum engineers	25%	771	193	840	210	902	225
22132	63	12	Safety engineers, except mining engineers	25%	469	117	539	135	606	152
22199	64	12	All other engineers	25%	5495	1374	6356	1589	7176	1794
22301	69	12	Architects, including landscape architects	30%	6155	1847	7016	2105	7825	2347
22309	72	12	Surveyors	40%	33527	13411	35863	14345	37934	15174
25102	76	12	Computer systems analysts, electronic data processing	40%	7471	2989	9966	3986	12669	5068
24305	78	12	Agricultural and food scientists	25%	185	46	200	50	214	53
24308	79	12	Biological scientists	25%	812	203	888	222	957	239

Table 6-2 *Wide Area, Public, Data Radio Opportunity (sample page Appendix F) (continued)*

The dramatic impact of this screening on 1991 jobs is shown in Figure 6-1.

all jobs	115,927,360
all mobile jobs	48,166,734
wide area mobile jobs	23,340,856
2 way radio wide area mobile jobs	8,903,592
public 2 way radio wide area mobile jobs	6,494,338
public data 2 way radio wide area mobile jobs	1,317,046

Figure 6-1 *Market Opportunity: 1991*

Remember, these are worker counts. The personal cellular phone user does not appear. Nor do these counts correlate directly to devices. It is quite possible for the public mobile data user to have multiple radios—even multiple data radios (railroad workers may have several wireless data terminals spotted throughout a yard). But this is an airtime service view. The worker may be as loaded with devices as a mule—that makes device manufacturers happy—but that same worker will only operate one device at a time.

Using this screening approach, and assuming a compound job growth of 1.5 percent per year (far greater than the past four years), only 1,437,732 two-way, public data jobs were seen for 1996. That included *both* the data-over-cellular and packet switched opportunities. Dividing this potential by the multiple alternatives that will be available in 1996 suggests difficult, competitive pricing with no large subscriber base for any regional offering.

A summary of the then-seen 1996 potential is shown in Figure 6-2. Note that the final bar groups several similar categories.

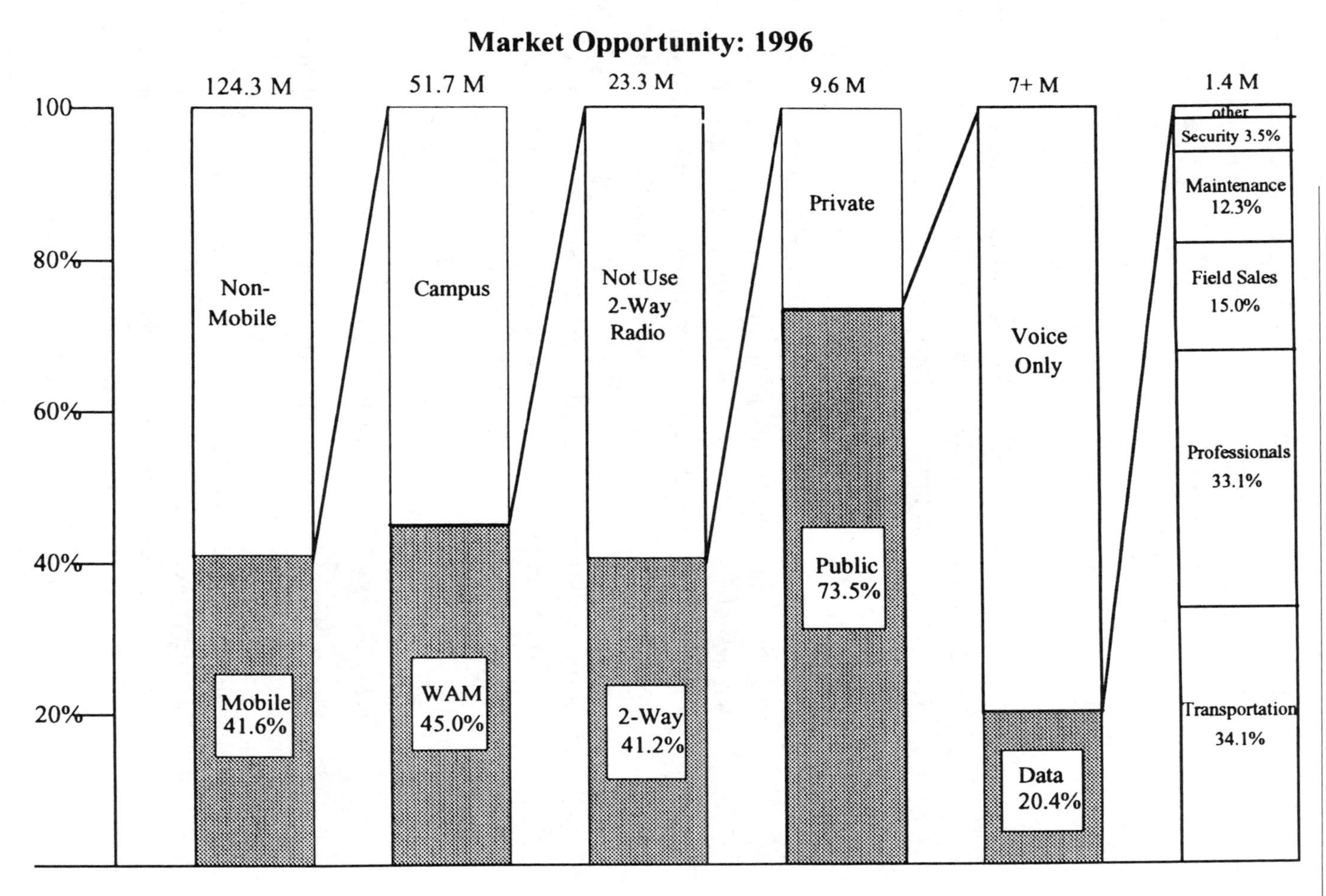

Figure 6-2 *Market Opportunity: 1996*

"Transportation" represented the largest public data service opportunity. Not surprisingly, nonmetro truckers are the target of Qualcomm's Omnitracs and Motorola's CoveragePLUS systems. Urban vehicles (messengers, bus drivers, couriers, taxis) are targeted by ARDIS (Red Arrow Messenger, UPS), RAM (Bell Ambulance, Coast Shuttle), and the SMRs, including RaCoNet.

"Maintenance," a much smaller opportunity, has already been well penetrated by ARDIS: ~15–18 percent of the opportunity. RAM is also working this territory (Hobart, National Computer).

Nearly half the opportunity—and still virtually untouched—came from two super categories: "Field Sales" and "Professionals." No single profile adequately describes these applications, but the users probably have by far the strongest need for laptops as opposed to specialized devices. Their information needs are probably greater also. E-mail should be routine for this class; RAM (and now ARDIS) is pressing very hard here for packet switched solutions. But there is also a marked requirement for facsimile and a residual need for voice communication. Thus, cellular will be a significant competitor for this segment.

6.2.2 Detailed Examples: 1991 Projection

The power of this job base organization can be demonstrated with the following example:

After the publication of the report one buyer asked the question: "Where, and how large, is the 1991 Public Data Radio Opportunity when the market is segmented so that only classifications having a job count greater than 10,000 are evaluated?"

An initial pass combined selected, fine-grain job breakdowns that, individually, would have been missed. For example, "Police and Detective Supervisors" and "Police Detectives and Investigators" were merged with "Police Patrol Officers"; "Brokers, Real Estate" were combined with "Sales Agents, Real Estate"; and so on.

The Public Data file was then sorted at the 10,000 job level producing the surprisingly compressed results shown in Table 6-3.

Table 6-3 *1991 Sort: Public WAM Data Candidates > 10,000: 1991*

BLS CODE	BLS SEQ	SEG	Description	% Data	# Public	# Data
			Totals:	30.6%	3559175	1087847
15026	6	11	Managers, food service and lodging	25%	44739	11185
19999	9	11	All other managers and administrators	25%	138387	34597
19005	16	11	General managers and top executives	25%	345337	86334
22309	72	12	Surveyors	40%	35863	14345
21114	25	13	Accountants and auditors	35%	92305	32307
32102	159	14	Physicians and surgeons	20%	268156	53631
32502	160	14	Registered nurses	20%	75722	15144
32505	198	14	Licensed practical nurses	20%	89691	17938
		15	"Reporters...writers... editors..."	35%	66327	23215
28108	110	19	Lawyers	15%	101446	15217
		20	"Special agents...insurance..."	25%	82442	20611
		20	"Brokers...real estate"	30%	160183	48055
49008	269	20	Non-tech sales representatives, exclude scientific and related products or services and retail	20%	243377	48675
43017	271	20	Sales agents, selected business services	20%	57669	11534

Table 6-3 *1991 Sort: Public WAM Data Candidates > 10,000: 1991 (continued)*

BLS CODE	BLS SEQ	SEG	Description	% Data	# Public	# Data
49005	274	20	Tech sales representatives, scientific and related products and services—except retail	20%	170421	34084
85705	568	41	Data processing equipment repairers	80%	39054	31243
85717	573	41	Electronics repairers, commercial and industrial equipment	80%	13743	10995
85726	574	41	Station installers and repairers, telephone	40%	38474	15390
85926	614	41	Office machine and cash register servicers	60%	35910	21546
85702	575	42	Television and cable TV line installers and repairers	30%	68773	20632
85110	577	43	Industrial machinery mechanics	30%	69204	20761
		49	"Automotive...bus...truck mechanics..."	13%	128582	16297
57311	316	51	Messengers	80%	13096	10477
97108	837	51	Bus drivers	30%	81887	24566
97117	841	51	Driver/sales workers	20%	127757	25551
97105	842	51	Truck drivers, light (includes delivery and route workers)	15%	330926	49639
97102	843	51	Truck drivers, heavy or tractor trailer	60%	553642	332185
		61	"Police...patrol officers"	51%	53726	27216
		62	"Fire fighters...supervisors"	45%	32335	14478

A second sort (Table 6-4) eliminated the BLS identifiers and distributed the jobs into the previously defined summary groupings. This improved organization permits one to focus on (and perhaps disagree with) key percentage assumptions.

Table 6-4 *1991 WAM Groupings*

Description	% Data	# Public	# Data
Totals:	30.6%	3559175	1087847
Management:	25%	528463	132116
Managers, food service and lodging	25%	44739	11185
All other managers and administrators	25%	138387	34597
General managers and top executives	25%	345337	86334
Professionals:	25%	972135	240463
Surveyors	40%	35863	14345
Accountants and auditors	35%	92305	32307
Physicians and surgeons	20%	268156	53631
Registered nurses	20%	75722	15144
Licensed practical nurses	20%	89691	17938
"Reporters...Writers...Editors..."	35%	66327	23215
Lawyers	15%	101446	15217
"Special agents...Insurance..."	25%	82442	20611
"Brokers...real estate"	30%	160183	48055
Field Sales:	20%	471466	94293
Non-tech sales representatives, exclude scientific and related products or services and retail	20%	243377	48675
Sales agents, selected business services	20%	57669	11534
Tech sales representatives, scientific and related products and service—except retail	20%	170421	34084

Table 6-4 *1991 WAM Groupings (continued)*

Description	% Data	# Public	# Data
Maintenance:	35%	393741	136864
Data processing equipment repairers	80%	39054	31243
Electronics repairers, commercial and industrial equipment	80%	13743	10995
Station installers and repairers, telephone	40%	38474	15390
Office machine and cash register servicers	60%	35910	21546
Television and cable TV line installers and repairers	30%	68773	20632
Industrial machinery mechanics	30%	69204	20761
"Automotive...bus...truck mechanics..."	13%	128582	16297
Transportation:	40%	1107309	442419
Messengers	80%	13096	10477
Bus drivers	30%	81887	24566
Driver/sales workers	20%	127757	25551
Truck drivers, light (includes delivery and route workers)	15%	330926	49639
Truck drivers, heavy or tractor trailer	60%	553642	332185
Security:	48%	86061	41693
"Police...patrol officers"	51%	53726	27216
"Fire fighters...supervisors"	45%	32335	14478

For example, it was assumed that 15 percent of the "Lawyers" who employed public two-way radio of any kind (almost exclusively voice cellular) were also candidates for data (laptops via cellular).

But 80 percent of all "Data Processing" service personnel using public two-way radio (ARDIS, RAM, SMRs, cellular) would be candidates for data. Given ARDIS's success in this target segment at YE91, the number has characteristics of a business forecast, not just an opportunity statement. It also suggested that ARDIS must branch into other application areas to continue their growth. This is clearly happening: Avis, Coca-Cola, G.O.D., UPS, and others.

The size of the "Truck drivers, heavy or tractor trailer" opportunity is obvious. Qualcomm can install Omnitracs at their present rate for 30 years and not run dry. CoveragePLUS clearly wishes to share in this opportunity and, with its RaCoNet alliance, is cranked up to share in both the airtime revenue and the device business.

Finally, the files were organized to reveal the reduction in numbers as each "test" was applied to the gross opportunity (Table 6-5).

Table 6-5 *Public Data Candidates: 1991 Sort*

Description	# WAM	# Radio	# Public	# Data
All Jobs Total:	23340856	8903592	6494338	1317046
Data Sorted Total:	9781350	4794362	3559175	1087847
Percent:	41.9%	53.8%	54.8%	82.6%
Management:	1436267	564370	528463	132116
Managers, food service and lodging	94188	47094	44739	11185
All other managers and administrators	615054	153764	138387	34597
General managers and top executives	727025	363512	345337	86334
Professionals:	2567500	1119827	972135	240463
Surveyors	44412	37751	35863	14345
Accountants and auditors	323877	97163	92305	32307
Physicians and surgeons	510773	357541	268156	53631
Registered nurses	336544	100963	75722	15144
Licensed practical nurses	472056	94411	89691	17938
"Reporters...writers...editors..."	111628	69818	66327	23215
Lawyers	213571	106785	101446	15217
"Special agents...insurance..."	209696	86781	82442	20611
"Brokers...real estate"	344943	168614	160183	48055

Table 6-5 *Public Data Candidates: 1991 Sort (continued)*

Description	# WAM	# Radio	# Public	# Data
Field Sales:	1182703	496280	471466	94293
Non-tech sales representatives, exclude scientific and related products or services and retail	731961	256186	243377	48675
Sales agents, selected business services	151759	60704	57669	11534
Tech sales representatives, scientific and related products and service—except retail	298983	179390	170421	34084
Maintenance:	858263	414464	393741	136864
Data processing equipment repairers	68516	41110	39054	31243
Electronics repairers, commercial and industrial equipment	24111	14467	13743	10995
Station installers and repairers, telephone	80998	40499	38474	15390
Office machine and cash register servicers	62999	37800	35910	21546
Television and cable TV line installers and repairers	180982	72393	68773	20632
Industrial machinery mechanics	132447	72846	69204	20761
"Automotive...bus...truck mechanics..."	308210	135349	128582	16297
Transportation:	2876010	1338814	1107309	442419
Messengers	72755	65479	13096	10477
Bus drivers	139978	90986	81887	24566
Driver/sales workers	283905	141953	127757	25551
Truck drivers, light (includes delivery and route workers)	995267	348343	330926	49639
Truck drivers, heavy or tractor trailer	1384105	692053	553642	332185
Security:	860607	860607	86061	41693
"Police...patrol officers"	537257	537257	53726	27216
"Fire fighters...supervisors"	323350	323350	32335	14478

Note that only 41.9 percent of all wide-area-mobile jobs fell into the 10,000 requirement, but 82.6 percent of the Public Data opportunity was covered in spite of this constraint. This bodes well for a "Top 500" marketing strategy.

This chart suggests other factors that might have directed a 1991 marketing thrust. The largest "Professional" categories are in the health-care field. Further, organizations exist that may make the marketing problem tractable. Appendix G contains three pages from InfoMed's Overview of Home Health Care. The number of home health agencies in the U.S. grew by 76 percent during the 1986–1991 period. The average cost per call in 1990 was $66, up 31 percent from 1985. Data radio could be a powerful technique for controlling these (largely) Medicare-funded costs. Note that InfoMed already has a development effort in place for pad-based PC devices.[11]

6.3 A NEW LOOK AT JOBS

The 1987 work, though usefully updated, cries out for a total revision. This is not exclusively a need for "fresher numbers"; job growth (or decline) moves slowly enough so that directions change at a relatively glacial pace. Nor is it only a desire to become somewhat less granular in job breakouts.

The best reason for a totally fresh update is to attack the job estimates by region so an individual city breakdown can be achieved.

This work is essentially complete. The goal—which has been achieved—was to estimate the opportunity by regions in which the resident and the job are coincident. A secondary goal was to pick up the "shadow economy"—those jobs that normally go unreported in labor statistics reported in the newspaper.

This extremely difficult task was achieved the following way:

1. Data diskettes were obtained from the Department of Commerce, Projections Branch, Regional Economic Analysis Division, in the Bureau of Economic Analysis (BEA). This raw summary data contained current statistics and projections on all coincident residents/jobs, including military, self-employed, and the shadow economy, broken down by 183 U.S. regions.
2. Separately, the Industry/Occupation Matrix (IOM) was obtained from the Bureau of Labor Statistics (BLS). This 15-year projection has 507 occupation classes, but does not include self-employed.
3. The BLS statistics were compressed from 507 to 236 occupation classes to provide a more tractable format.
4. Self-employed estimates were added to the BLS statistics based on current census data. There were also a series of normalcy checks (for example, farmers have high self-employed ratios; state/local/federal governments are near zero) to ensure that the addition was valid.
5. The IOM and the BEA statistics were then forced onto the same industry base. During this process all BEA military jobs were excluded.
6. This forced merger revealed two industry problem areas: public hospitals and education are placed under state/local government by the BEA. The IOM tallies them under service industries.
7. The problem areas were handled by creating two unique industry classes:
   a. Hospital workers—federal bulletins assisted in the manual distribution, with high confidence.
   b. Education—distribution was more agonizing, with cross checks on both state capitals and "university cities." Although there is somewhat less confidence in the precision of this allocation, it is clearly directionally correct (mismatch <10 percent).
8. The base year, 1993, was created by calculation from the 1991 and 1995 estimates.

The result is the ability to tailor a detailed statistical analysis by *region* (from New York City to, say, Aberdeen, SD) in four years of interest (1993, 1995, 2000, 2005) by 14 industry classes and 236 invidual occupations.

A tantalizing extension, which could be attempted in cooperation with ARDIS's copyrighted Network Comparison Matrix, is not only to allocate the opportunity by region down to two-way public data radio, but also to estimate on which network that job best fits. This distinction is probably achievable for at least the data-over-cellular versus packet switched categories.

REFERENCES

[1] *Computer Age—EDP Weekly*, 4-28-93.
[2] *Data Communications*, February 1993.
[3] *Cellular Sales & Marketing*, February 1993.
[4] ARDIS Lexington Conference, 7-27-93.
[5] Ibid.
[6] Ibid.
[7] Ibid.
[8] *Mobile Communications Market: A Business Consumer View*, 1987 JANUS Group of Connecticut.
[9] As reported in the *New York Times*, 9-27-92.
[10] Dennis F. Strigl, BAM President/CEO, as reported in the consortium press release, 5-7-92.
[11] JFD 1993 discussions with Fred Neufeld, President of InfoMed.

CHAPTER

7

CONVERGENCE AND CONFLICTS

7.1 HARMONY HELPS

Pricing tugs-of-war can pit service providers against users. But another series of goals pursued by both parties can harmonize differences—or deepen the conflict. At the highest level the network service provider makes investment decisions based on:

1. Subscriber loading, usually measured by active-user-per-base station capacity
2. Schedule rollout: how long before the infrastructure can be operational
3. Infrastructure versus spectrum tradeoffs required to achieve acceptable loading

Meanwhile, the user makes buy decisions based upon three different factors:

1. Coverage
2. Response time
3. System availability

Some service provider investment decisions, such as increasing capacity by adding base stations, improve the user's coverage; in certain cases they can also improve response time—serendipity!

Other service provider decisions, such as high(er) bit rates, receive much marketing emphasis but are not usually noticed by the user because the extra capacity generated is, or will be, consumed by additional users—neutrality.

Finally, a service provider's rush to bring a system online quickly can be harmful to the user's desire for high availability—potential enmity.

7.2 SERENDIPITY

7.2.1 Subscriber Loading

Subscriber loading is a (too) loose way to characterize load. In any network "slippage"—the ratio of active users to subscribers—is always present. At the most mundane level, a dedicated dispatch workforce always has some members ill, on vacation, in training, or on second shift so that the most efficient ratios rarely exceed ~80 percent of subscribers active in a busy hour. Still, describing the load in terms of field service subscribers is a rough and ready way to grasp the load.

For inquiry applications, such as to a NewsNet database, it is rare to find more than 70 persons active from a subscriber pool of tens of thousands. Clearly what one needs to know here is the number of active users.

Active user loading characteristics depend upon the application type. Key determinants include:

1. Message length, which sometimes can be shortened by techniques such as data compression
2. Message rate—the interval between successive messages
3. The speed of the target—with very noticeable differences occurring between fixed location transmissions and a vehicle traveling at 60 miles per hour

Active users in an area are, of course, distributed across multiple base stations. This distribution is never uniform; one cannot achieve a realistic capacity estimate by multiplying single-station potential times the number of stations in the area. In small groupings (~30 base stations) it has long been known[1] that a single cell may carry 3 to 4 times the load of the average cell. Still, useful load dispersion does occur.

As bit rates rise, data packet time intervals tend to be very short: A busy individual might only consume 1 to 1.5 seconds of pure inbound transmission time during the busy hour (the network will use considerably more time to deliver the message).

Thus, in an imaginary 30-base-station metro area one might find 10,000 field service subscribers accommodated as follows:

Monthly subscribers	10,000
Active users, busy hour (80% of subscribers)	8,000
Worst cell load, busy hour ((8000/30) × 3)	800
Inbound seconds used, busy hour (1.25 × 800)	1,000

Thus the service provider focuses on multiple, well-sited base stations to disperse the load, and favors high bit rates (assuming reasonable efficiency) to shorten the transaction interval. These actions raise the active user capacity and, thus, the revenue potential of the network. But additional base stations are also good for the user because they usually translate to improved coverage.

7.2.2 User Coverage

Coverage needs are application specific. Some users (taxis) can be satisfied with street-level coverage targeted to their route areas. Other users (high-tech field service) require in-building penetration. Irritating gaps exist in the best of systems and the pressure for better coverage is ever present. Circuit switched cellular, with its relentless, continuous investment in infrastructure, including many rural areas, offers the best long-term prospects for ubiquitous wide-area metro coverage.

A trivial example: PacTel Cellular began service in Los Angeles in June 1984 with 16 base stations, less than half the stations then in place on IBM's private DCS system. By March 1986 PacTel had 31 stations in operation, about the same as DCS; by YE86 there were more than 60;[2] about four times that number existed at YE92; and there may be 300 by YE93. And PacTel is not the only service provider. LA Cellular became operational in March 1987 with 40 base stations and has followed its own sharp growth path. ARDIS has also grown post-DCS, but not at the staggering cellular pace.

Public packet switched systems have four service alternatives to circuit switched cellular:

1. Offer only street-level coverage. This is done by design in CoveragePLUS and RaCoNet.
2. Provide cellular-like, targeted coverage. The best example is RAM. No attempt is made to match cellular base-station-for-base-station. Instead a new layout, with its

own reuse constraints, is built in areas of high business activity. Building penetration can range from undesirable to excellent.

3. Provide overlapping, receive-multiple-copy base stations with the specific goal of improving building penetration. The best example is ARDIS. Where it chooses to compete, this approach will generally provide better in-building coverage. Empirical evidence of this probability is that ARDIS contractually guarantees its coverage contours.
4. Match the voice cellular infrastructure, base for base, as is the design goal of CDPD. If this truly happens, CDPD will have important wide-area coverage. Its then many (~8 times that of ARDIS) small cells may improve in-building penetration to the point of academic argument. The jury is out.

As always, it is possible to have variations on these approaches. Bell-ARDIS (Canada) has both handheld, low-power deep penetration and vehicular, high-power, street-level systems. Each operates on separate frequencies as shown in Appendix H.

If in-building penetration is not a key issue for a specific application, ARDIS has no coverage advantage. Indeed, its concentrated focus (even more intense with RAM) tends to exclude areas of reduced business activity. Only about half the counties of California, for example, have an ARDIS base station—not optimum for a utility company ranging the backwoods.

7.3 NEUTRALITY

7.3.1 User Response Time Problems

The first thought is that data-over-cellular has a response time disadvantage because of its lengthy dial setup time. But much of this time can be masked from the user. If a parts order entry form is prepared on a laptop and <ENTER> is pressed the user can proceed to other tasks while the PC places the call, enters the user ID and password, and transmits the message. Further, once connected the user "owns" the channel. There is no inbound contention, no queued outbound responses.

After the initial daily sign-on, a packet switched system often requires no further user attention (there are exceptions). But each time the unit is powered on a 30-second delay can occur until the modem finds the base station. This is roughly equivalent to cellular dial time.

Further, inbound messages must contend for channel resources with an ever-changing peer group. When the <ENTER> key is pressed the subscriber unit will (typically) listen for a busy indication. If someone else is transmitting, the message waits till the channel is free. And if the channel does seem free, it is still possible for a collision to occur, forcing retransmission of both messages after a random time-out delay.

Further, a significant portion (depending upon signal strength, target speed, etc.) of the contention-free messages are damaged in transit and require retransmission. A 2- to 3-second time-out waiting for an acknowledgment can make inbound packet entries sluggish.

On the outbound side the messages must be queued at the base station for a free channel. If the base is working on a full-screen display response, the wait times can be noticeable.

Improving bit rates, though certainly welcome, is rarely decisive for short messages. For user messages 100 octets (800 bits) in length the transmission time difference between a vintage protocol (4800 bps/40 percent efficient) versus a modern protocol (19,200 bps/60 percent efficient) is less than 350 milliseconds. Measurable—but not necessarily noticeable. And if the modern protocol is employed to bring many more users to the system the increased contention can quickly offset some of the difference.

There are many stories (some apocryphal) of horrible packet switched response time delays, with multimessage transactions taking twice as long as their landline equivalents.[3] The average network response time on today's (vintage) ARDIS system is 4 to 5 seconds for a roundtrip message; ARDIS guarantees a response time of 10 seconds.[4] RAM's current subscriber load is too low to reveal response time problems. But packet switched radio, in general, is not lightning fast.

7.3.2 Coverage Solutions that Improve Response Time

RAM deploys its infrastructure in a cellular-like configuration. Adjacent base stations operate on unique frequencies that are reused only at a noninterfering distance. If two users happen to compete simultaneously for the same base station they will be randomly rescheduled after the collision.

For coverage reasons ARDIS employs its infrastructure so all base stations are tuned to the same frequency. By design each inbound transmission is heard by multiple base stations. If two users happen to compete simultaneously, there is a high probability that the message will be heard on one (or more) adjacent base stations and salvaged. This reduction of inbound contention reduces the random waits associated with rescheduling and can improve response time. This is illustrated in Figure 7-1.

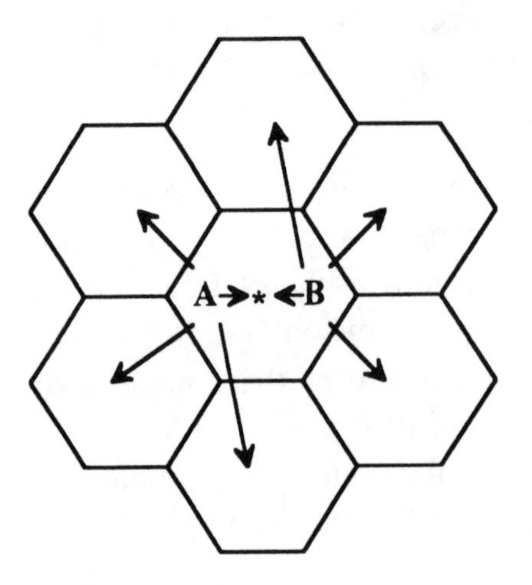

A & B collide in the cell in which they reside, but are "heard" by adjacent cells tuned to the same frequency.

Figure 7-1 *ARDIS Inbound Contention Approach*

7.3.3 Infrastructure versus Spectrum

The infrastructure versus spectrum tradeoff is complex. In theory, packet switched data will offer far larger revenue per channel than voice or data-over-cellular. Assuming $.38/minute for voice and a channel used 65 percent of the time in a busy hour, the best-case voice revenue per channel per cell per hour is $14.82. With a packet price of $.0128/12 octets (see Chapter 3) and a message profile of 72 octets in and a 120-octet response, only ~72 transactions in an hour, one every 50 seconds, yields the same revenue.

But voice revenues are real; data has been "just around the corner" for 20 years. A carrier cannot reasonably be expected to sacrifice multiple revenue-producing voice channels for a potential data opportunity. The opportunity cost is too painfully evident.

Thus, a range of potential vendor solutions exist. At the two extremes lie Motorola RD-LAP alternatives versus CDPD. In bluntest form the differences between the two approaches are these:

1. RD-LAP provides a uniform (theoretically lower) cost infrastructure that delivers more capacity if more spectrum is allocated to it during the design phase. A high-capacity configuration requires 10 or more dedicated *cellular* channels (yield: 12 RD-LAP channels); the lowest capacity configuration (ARDIS) would require only one. But the infrastructure investment for either extreme is identical.
2. CDPD requires a rich (possibly costlier) infrastructure investment to achieve very high capacity—but at no cost in spectrum.

Each approach has its own particular set of problems. The important point is that the service provider must constantly think through the tradeoff required between two precious resources and the user couldn't care less as long as the message gets through.

7.4 POTENTIAL ENMITY

Wide-area deployment is a time-consuming process under the best of conditions. If the system is new or untried, testing intervals quickly expand to multiyear endeavors. Figure 7-2 is a graphic portrayal of key milestones on three Motorola implementations.

The infrastructure of the first system, IBM's DCS, was a modification of a prior development effort. From contract to reasonable stability (citywide failure about once a week) required 3 years of dedicated work; to completion of the nationwide rollout required about 4.5 years.

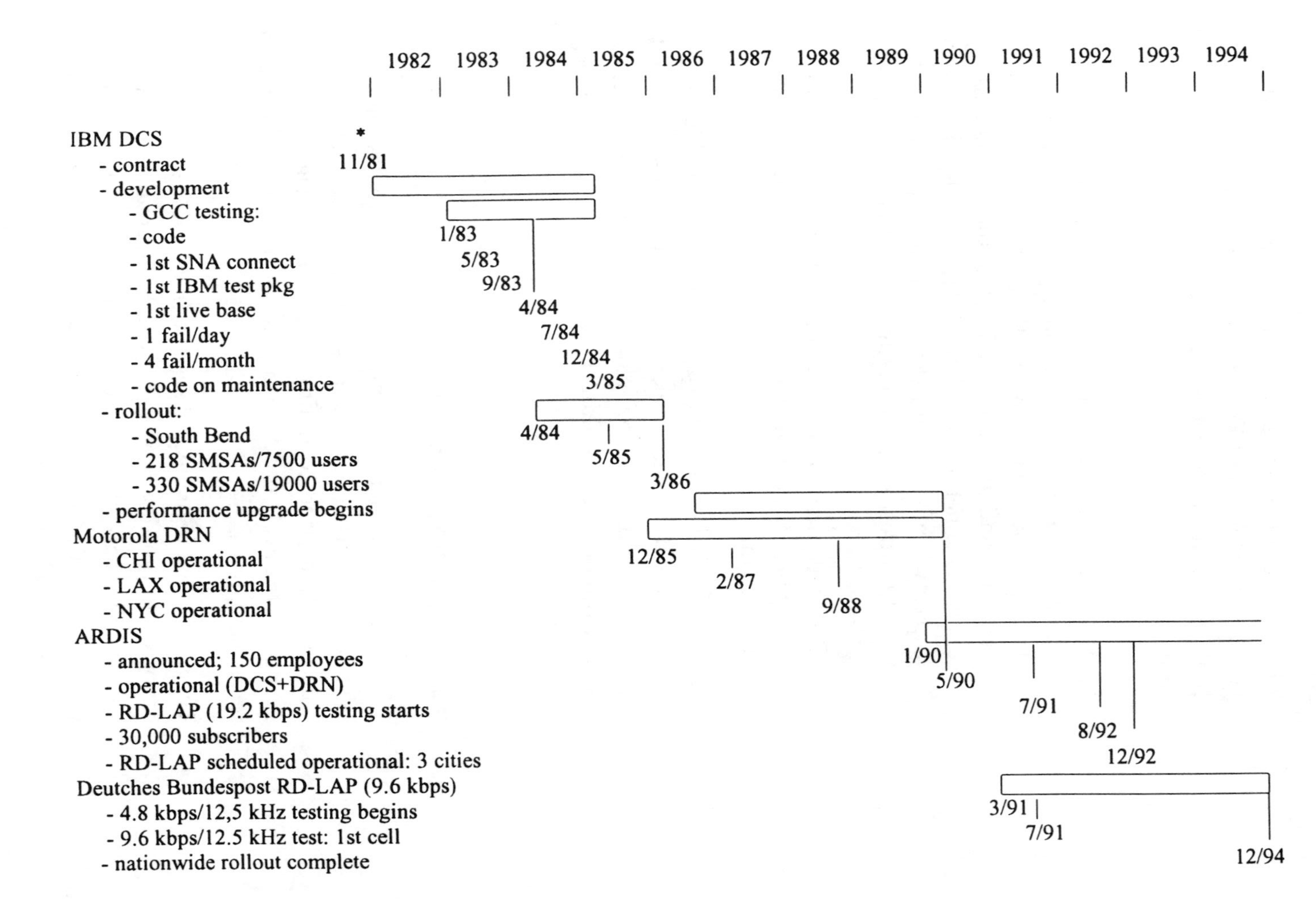

Figure 7-2 *Representative Motorola Schedule Rollouts*

The last system, Deutches Bundespost, began testing slow-speed versions of RD-LAP in March 1991. Key elements of this system were derived from design done for Motorola's own earlier DRN configuration. National rollout is scheduled to be complete 3.75 years after testing began.

The RD-LAP variation used in the ARDIS upgrade began its test in July 1991. The first city was operational 18 months later. Ironically, production use has been blunted by Motorola's failure to deliver dual protocol modems on schedule. Nearly 2.5 years will elapse before authentic subscribers roaming between Washington and Baltimore can use the system (ARDIS has had to tune the existing base to achieve the needed 20 percent extra capacity). Rollout of the top 30 MSAs is expected to require 4 years.

These are not unusual intervals. RAM, employing a high-speed version of the fully functional Mobitex infrastructure, achieved 200 operational MSAs about 4 years after contract award.

ARDIS has achieved a consistently high level of availability (99.97 percent) after years of effort (RAM's achievements are not known). The CDPD drive to place new users on networks still in the test phase is crucial from the service provider's viewpoint: users = revenue. But early users will inevitably slam into reality. Systems will fail; painful outages will occur. Thus, the new user must resist jumping too quickly. Long periods of pilot testing must precede full-scale commitment, leading to conflict between participants.

REFERENCES

[1] CTIS/Telocator Report (prepared by Compucon), "An Analysis of Spectrum Requirements of the Cellular Industry," 1985.

[2] Reed Royalty, PacTel Vice-President of External Affairs, as reported in *Telocator*, June 1986.

[3] Barry McDonald, HP Custom Service Manager, as reported in *En Route Technology*, 2-19-92.

[4] Matthew Whitehead, ARDIS Counsel, Lexington Conference, 7-27-93.

CHAPTER 8

AIRTIME PRICE PROJECTIONS

8.1 GREAT EXPECTATIONS

"When Bell Atlantic Mobile Systems (BAMS) launches its first . . . CDPD . . . system in Washington . . . it will initiate a new business that will impact the . . . bottom line by 1995 . . . . The average monthly bill for cellular is dropping, and . . . new services such as data will boost those bills."[1]

Can data meet the high revenue expectations of the many service providers bringing additional capacity to market? Note that ARDIS current infrastructure is perhaps 3 to 4 percent loaded (see Chapter 5). RAM has huge untapped capacity. ARDIS is in the midst of a major infrastructure upgrade that will quadruple its existing capacity—and have even more dramatic impact in the largest cities. CDPD, if rolled out to all cellular base stations, has enormous capacity potential. What does this mean for CDPD's airtime price outlook?

8.2 SOME REASONABLE ASSUMPTIONS

To probe at the pricing limits one must make a series of assumptions. For simplicity of calculation, some of the assumptions brace data-over-cellular with (perhaps) excessive vigor. The key assumption is that the carriers do not wish to drive their revenue down. Thus, data will be priced to yield at least revenue parity with voice. Carriers are assumed to be the price drivers; ARDIS and RAM will come into line if falling CDPD prices attract customers.

Specific assumptions include:

Circuit switched cellular price per minute (1993)	\$.60
Circuit switched cellular price per quarter minute (1994)	\$.15
Cellular billable dial time	30 seconds
Cellular billable handshake time	15 seconds
Cellular fax modem speed (1993):	4800 bps
Cellular fax modem speed (1994):	7200 bps
Facsimile information:	
a - single 8″ × 10″ page, Group III, 203 × 98 dpi	
b - resulting ~1.6 million bits Huffman coded to:	160,000 bits
c - bits can be packetized to:	20,000 octets
User octets in a CDPD packet	120 octets

8.3 FACSIMILE

8.3.1 Circuit Switched Cellular

Today's price to send a single page is $1.20 calculated as follows:

Dial time		30.0 seconds
Handshake time		15.0 seconds
Transmit time =	$\frac{160{,}000 \text{ bits}}{4800 \text{ bps}}$ =	33.3 seconds
		78.3 seconds

At 1-minute granularity, the 78-second transaction time converts to 2 minutes × $.60 = $1.20.

A few cellular fax machines (Powertek) already transmit at 7200 bps. Assuming widespread achievement of this speed, and a controlled step to subminute billing, the price to send a single page will drop to $.75 calculated as follows:

Dial time		30.0 seconds
Handshake time		15.0 seconds
Transmit time =	$\frac{160{,}000 \text{ bits}}{7200 \text{ bps}}$ =	22.2 seconds
		67.2 seconds

Five quarter-minute intervals × $.15 = $.75. A change to per-second billing could tighten the price to $.68 but the coarser projection is adequate for this analysis.

8.3.2 CDPD Packet Transmission

The May 11 CDPD Version 0.9 press release, page 2, states: "the CDPD technology will . . . enabl[e] one portable computing device to offer . . . phone, . . . fax." One must take care in interpreting these press releases. For example, CDPD will not transmit voice over the packet network. The "phone" reference probably means that the cellular radio in the laptop can be employed for voice using a handset, or data in CDPD packet mode.

Similarly, true facsimile could be sent via circuit switched cellular. Or a "fax" message could be just the text portion sent via the packet network to a fax server for the addition of letterheads and conversion to facsimile output. However, assume for now that the goal is to send a compressed image via CDPD to a fax server—and to do it at a price that is competitive (but not less than) circuit switched cellular.

Because the cellular fax may well have transmission errors present, assume that the packet transmission will be sent as a point-to-point, unacknowledged, multiframe message.[2] The frames are unnumbered; there will be no flow control.[3] (More is provided on this technical jargon in later technical chapters.) If errors occur they will be ignored.

A problem with the unacknowledged message is that header compression cannot be employed.[4] For this analysis that capacity-limiting problem will be ignored. A compressed header permitting a four-block packet to carry 120 octets is assumed.

Thus, CDPD can theoretically deliver 20,000 outbound octets in <15 seconds calculated as follows:

$$\frac{20{,}000 \text{ octet message}}{120 \text{ octets per packet}} = \begin{array}{l}167 \text{ packets} \times 4 \text{ blocks/packet} \times \\ 420 \text{ bits/block} = 280{,}000 \text{ bits}\end{array}$$

and

$$\frac{280{,}000 \text{ bits}}{19{,}200 \text{ bps}} = 14.58 \text{ seconds}$$

Assuming 100 percent utilization of the channel a transmitted page could be priced at $.15 and still yield the same revenue as a voice-equivalent channel. The octet price would be:

$$\frac{20{,}000}{\$.15} = \$.10 \text{ per } 1{,}333 \text{ octets}$$

To inject some sanity into this calculation assume that the channel is only 75 percent utilized (or a 25 percent contingency is applied). That yields the convenient calculation of $.01 per 1,000 octets, or $.001 per 100 octets.

8.4 INTERACTIVE BALANCED LOAD

Assume the packet network is being used in a "traditional" way with interactive, bursty, short-message traffic and that this load is well balanced, meaning the message profile does not unduly force one of the channel pairs into unused idleness. If the average length of the outbound message is 200 ± 15 octets and the inbound message is 75 ± 15 octets then a message rate of 3.1 messages/second is sustainable[5] with both errors and channel hopping.

Now apply the packet price of $.001 per 1,000 octets. The revenue yield is:

Outbound:	.002 × 3.1 msg/sec × 60 sec =	$.372
Inbound:	.001 × 3.1 msg/sec × 60 sec =	.186
		$.558 per minute

This is very close to the voice yield of $.60 per minute and probably represents the low price case. If for any reason perfection is lost—say if the load is unbalanced—the revenue yield slides. Further, if outbound message lengths increase sharply, and selective ARQ is faithfully performed, the inbound channel revenue suffers from the necessary increase in the "selective reject" (SREJ) commands.

8.5 MIXED INTERACTIVE/"FACSIMILE"

If long image messages are introduced in busy channels carrying interactive traffic they should be segmented and sent on a delayed basis. The introduction of an additional 15-second response delay in an interactive sequence is probably unacceptable. Thus at peak intervals an outbound "fax" might be delayed for minutes waiting for idle time in the queue. A less efficient inbound "fax" will produce heavy contention that simply must be tolerated.

8.6 MOST LIKELY PRICE SCENARIO

In the absence of a uniform accounting and billing system, and with zero capacity constraints, the initial CDPD system is almost certain to be a low flat rate—perhaps $40 per month with multi-user discounts. Traffic profiles will be monitored to establish length characteristics.

Assuming no profile surprises, a $15 per month subscriber fee and a $.01 per 100-octet usage rate will be introduced. This usage rate is one-half RAM's current incremental rate (and far less than ARDIS). At the prior user-conditioned $40 per month level the $25 residual usage fees will purchase ~100 packets per day. In an e-mail environment this is equivalent to receiving ~4, 2,000-octet, single pages per day, along with the necessary e-mail inquiry costs, sign-on, etc. The claim can be made that the four letters cost less than the U.S. Post Office (approximately true). This will not yield a competitive "fax" price ($2) but that market can be captured on the circuit switched side.

As competitors respond prices will be driven downward to ~$.005 per 100 octets, one-fourth the current RAM price. At the $.005 level a *balanced* interactive load will yield nearly four times the revenue now achieved by voice cellular:

Outbound:	$.008 \times 3.1$ msg/sec $\times 60$ sec =	$1.49
Inbound:	$.004 \times 3.1$ msg/sec $\times 60$ sec =	.74
		$2.23 per minute

This is a comfortable margin. A balanced configuration is not crucial because there is so much maneuvering room. Fax will price out at $1.00—midway between current and projected cellular capability—but slower to deliver.

In the short term—1995 or so—one should be able to plan on a price of ~$.005 per 100 octets. Will this level "boost those monthly bills"? Not necessarily.

REFERENCES

[1] Benjamin L. Scott, Executive Vice-President and Chief Executive Officer, BAMS, *Telephone Week*, 6-21-93.
[2] CDPD V0.9, p. 403.4, section 2.1.1.
[3] CDPD V0.9, p. 403.4, section 2.2.1.
[4] CDPD V0.9, p. 403.4, section 2.2.
[5] JFD Associates, CDPD V0.8 Airlink Capacity Analysis.

PART

TECHNICAL

CHAPTER

9

AIRTIME PROTOCOLS

9.1 THE PACKET REVISITED

A "packet" is the information unit formed when a message is partitioned into more manageable sections for transmission and recovery. Most landline packets have three logical subsections formed when control information is added to user data (see Figure 9-1).

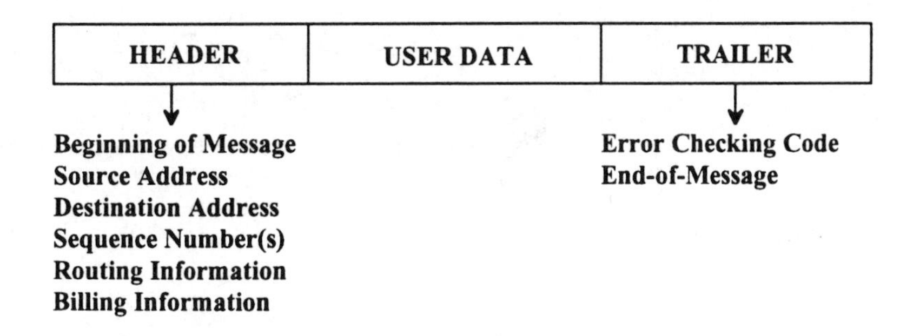

Figure 9-1 *The Packet*

Packets are not required to travel only on packet switched networks. They function quite nicely on circuit switched systems —often a source of confusion for persons encountering "adaptive packet assembly" in Microcom's MNP class 4 protocols.

To grasp the principle of the packet, the most common form of landline implementation will be discussed first. It is possible to transmit landline packets over radio links, as is routinely done in data over cellular. As bit rates rise, however, error-*checking* codes become less satisfactory and are usually augmented with error-*correction* codes. These differences will be explored second.

9.1.1 Message Segmentation: The Flag

When the user message is partitioned into packets a reserved separator character called the *flag* marks both the beginning and the end of the packet (see Figure 9-2).

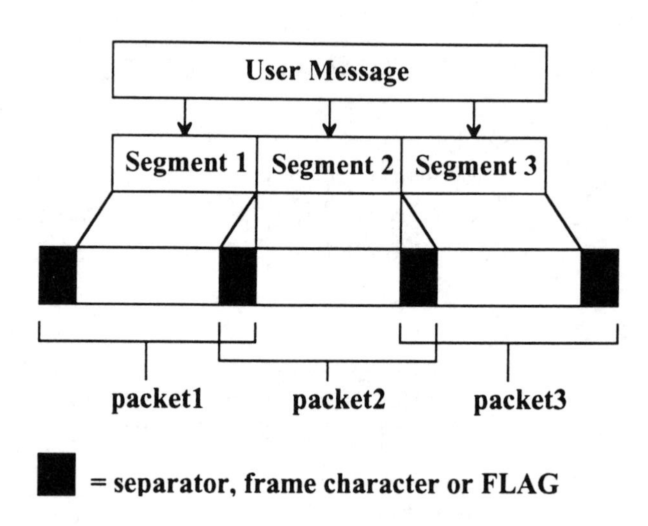

Figure 9-2 *The Flag*

The most common flag implementation, found in such diverse protocols as SDLC/HDLC, MNP, and CDPD, is the 8-bit sequence: **01111110**.

In these protocols the flag is one of three reserved patterns. The flag can also be manipulated in hardware by a technique known as "bit stuffing," which permits data transparency. These special variations will be discussed shortly.

9.1.2 The Address Field

The field following the flag is the address. Its minimum size is 8 bits, which yields only 256 unique addresses. In practice, address fields are *very* much larger, leading to the use of complex surrogate techniques to reduce the associated transmission overhead. One common technique is to uniquely identify source or destination address depending upon the function being performed. Using balanced address rules (unbalanced rules also exist) the field specifies the destination for commands; the same field specifies the source for responses (see Figure 9-3).

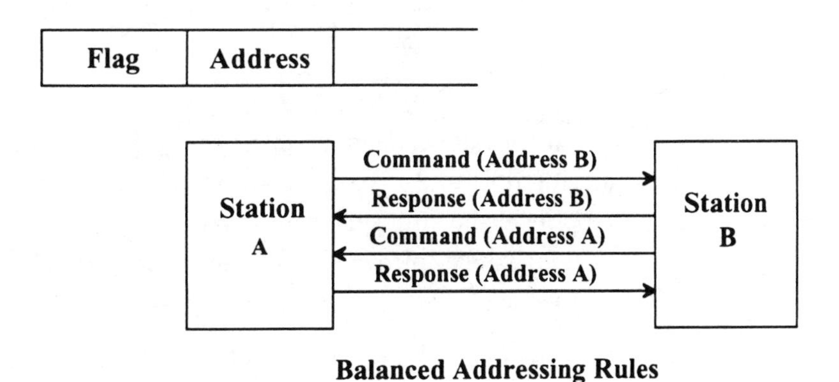

Figure 9-3 *The Address Field*

9.1.3 The Control Field

Following the address is a tightly packed field that defines the functions to be performed (see Figure 9-4).

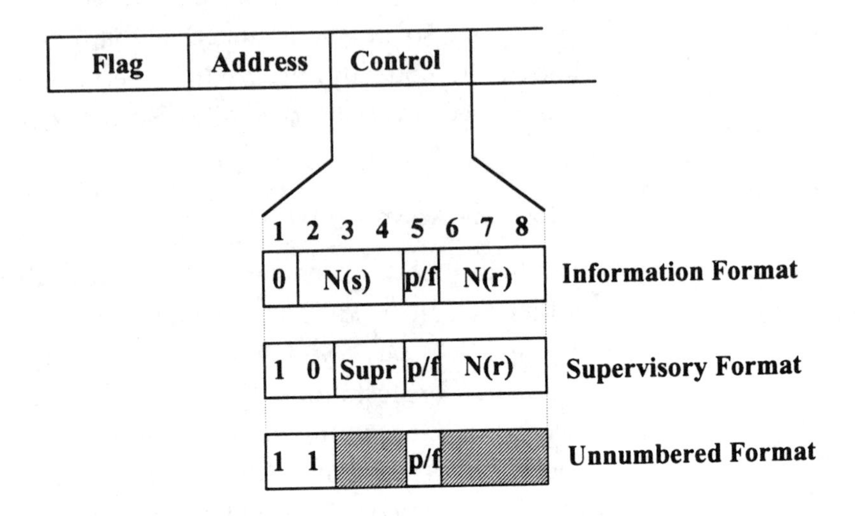

Figure 9-4 *The Control Field*

The information format controls the transmission and acknowledgment of end-user data, along with a poll to solicit a status response; the supervisory format controls requests for retransmission; the unnumbered format controls initialization and disconnection.

In the single-octet implementation there is room enough for the information format to track the sequence numbers of seven separate packets on both the send and receive sides. When the inbound response is unpredictable, as in CDPD, this field is expanded so 128 packets may be outstanding in each direction.[1]

9.1.4 The Information Field

User data is optional and is only associated with the information format. The information field follows the control field, can be variable length up to a defined system limit (typical defaults are 128–256 octets), and is "transparent": any format is legal—which puts user data into potential conflict with the flag and the other two reserved patterns. This problem is solved by "bit stuffing" as shown in Figure 9-5.

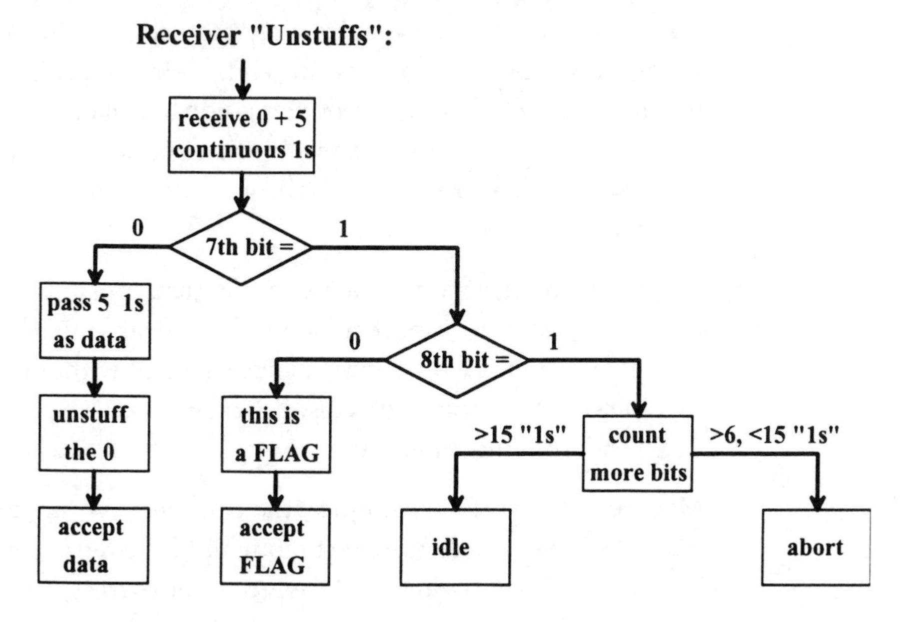

Figure 9-5 *The Information Field*

9.1.5 The Frame-Check Sequence Field

The final field, occurring just before the ending flag, is the frame-check sequence. Usually called the CRC, or cyclic redundancy check, this field is used for error *detection*. In its most common form the CRC is two octets long, quite useful for lower speed (~2400 bps) transmission. As transmission speeds rise this simple code can let many undetected errors slip past.

Modern implementations such as those employed by Motorola in RD-LAP use four-octet CRCs. Naturally there is an overhead penalty, but the undetected error rate plummets. These trade-offs, as well as the use of error *correction*, will be discussed next.

9.2 ERROR-HANDLING APPROACHES

9.2.1 Philosophy

A key distinguishing characteristic between airtime protocols is their particular philosophy of error handling. Nearly all vendors claim to have error-correction protocols. In the strictest sense, however, many do not. Instead, their protocols actually feature error detection and retransmission to (finally) produce a clean copy. A well-known example of a vendor that claims this form of error correction is Microcom with its MNP "defacto standards."

A true error-correction protocol is one that transmits enough redundant data, in a particular mathematical way, to permit faulty information to be corrected upon receipt without retransmission. All Motorola protocols, Ericcson's Mobitex, as well as CDPD fall into this category.

Many protocols mix techniques; the combinations are endless. It is possible to use error-detect mode until the retransmissions become onerous, switch to forward error-correct mode, and back that up with yet another level of detect/retransmit. All protocols, detection or correction, have breaking points. The power of the error-correction technique is usually expressed as the UBER, the undetected bit error rate. This is the frequency with which a bit in error escapes unnoticed into the data processing system. The less often this happens, the better—but good UBERs have a practical cost that may not be justified in some applications.

9.2.2 Error Detection versus Correction Basics

Assume an artificial world in which no more than 1 bit in 100 would ever be in error (there are such environments, though not in data radio). The protocol designer might then enforce a maximum block size of 10 EBCDIC octets. If the single-bit error was not frequently encountered, the designer might settle on a detect/retransmit scheme. Each octet could be assigned, say, a single odd parity bit. The detection of the single-bit error in the transmitted letter "d" could be represented as shown in Figure 9-6. Thus, 1 bit in 9 is redundant, yielding a "pure" protocol efficiency of ~89 percent.

Letter	EBCDIC	Parity	Bits sent	Bits received
u	10100100	+ 0 =	101001000	101001000
n	10010101	+ 0 =	100101010	100101010
d	10000100	+ 1 =	100001001	101001001 <= even parity
e	10000101	+ 0 =	100001010	100001010
r	10011001	+ 1 =	100110011	100110011
w	10100110	+ 1 =	101001101	101001101
a	10000001	+ 1 =	100000011	100000011
t	10100011	+ 1 =	101000111	101000111
e	10000101	+ 0 =	100001010	100001010
r	10011001	+ 1 =	100110011	100110011

Figure 9-6 *Error Detection: Parity*

Assume the single bit in 100 rule holds, but that the occurrence is unpleasantly frequent (at least at times). In the worst case every block would be contaminated. Retransmission could occur endlessly, but no useful information would ever reach the destination. The designer could now elect to add forward error correction to the block by adding an LRC, or vertical parity, code.

The transmitted block would now look like Figure 9-7.

Letter	EBCDIC	Parity	Bits sent	Bits received	
u	10100100	+ 0 =	101001000	101001000	
n	10010101	+ 0 =	100101010	100101010	
d	10000100	+ 1 =	100001001	101001001	<=even parity
e	10000101	+ 0 =	100001010	100001010	
r	10011001	+ 1 =	100110011	100110011	
w	10100110	+ 1 =	101001101	101001101	
a	10000001	+ 1 =	100000011	100000011	
t	10100011	+ 1 =	101000111	101000111	
e	10000101	+ 0 =	100001010	100001010	
r	10011001	+ 1 =	100110011	100110011	
(lrc)	11001110	+ 1 =	110011101	110011101	
				↑ even lrc	

Figure 9-7 *Error-Correction Example: Parity and LRC*

Thus 18 bits in every 99 are redundant, for a "pure" protocol efficiency of ~81 percent. But retransmissions have been stopped because the location of the error is known—caught in the crossfire of the two parities. The designer tradeoff is thus reasonably straightforward. If only a few blocks in 10 will have an error, use error detect/retransmit. If many blocks, including all 10, will contain an error, expend the overhead and use forward error correction. For a 100-octet message with exactly one bit error every 10 blocks this can be demonstrated as follows:

	User octets	Bits/ block	Numbr blocks	Bits sent	Bits resent	TRIB efficiency
Detect	100	90	10	900	90	800/990 = 80.8%
Correct	100	99	10	990	—	800/990 = 80.8%

Although real-world scenarios are fiercely more complex than this trivial drill, it is exactly this kind of tradeoff, on a far grander scale, that eternally faces the protocol designer.

9.2.3 Error Detection versus Correction: Vendor Examples

In practice, bit errors rarely occur in isolation, a phenomonon associated with miscellaneous transmission impairments such as Gaussian noise. These scattered errors are ever present, of course, but are accompanied by troublesome bursts associated with Rayleigh fading (and other problems). This error profile makes simple parity techniques quite impractical.

There are four general methods of attacking real-world problems, all of which are combined with ARQ (retransmission) techniques in different ways by different vendors:

1. *Error detection via CRC:* The most common CRC is the 16-bit CCITT V.41 generator, which detects all possible single-error bursts not exceeding 16 bits, and 99.9984 percent of all possible longer bursts.[2]

   When uncompressed bit rates are low, such as 2400 bps as used in Millicom's CDLC and many Microcom MNP devices[3] on cellular, burst errors of 16/2400 = 6.7 milliseconds can always be detected. For reference, note that the AMPS design assumes 2.5 ms fades at 20 mph.[4]

   As bit rates rise the transmission becomes vulnerable to long burst errors. Thus the 32-bit CRC has begun to appear in some protocols. This approach detects all bursts of 32 bits or less, 99.99999995 percent of bursts of length 33, and 99.99999998 percent of bursts longer than 33 bits[5].

2. *Weak error correction; interleaving; CRC:* A simple, low-cost, fast response—but weak and inefficient—error-correction technique is employed. For a fixed data block, one- or two-bit errors are *always* correctable. The block's bits are interleaved prior to transmission; if a burst error

causes damage, the errors are thus scattered upon reassembly; they are then attacked by the weak FEC.

This approach is employed in Ericsson's Mobitex (RAM) and Motorola's MDC4800 (ARDIS) systems. The Motorola example is representative: a rate one-half (50 percent of bits transmitted are "parity"), $k = 7$ convolutional encoding algorithm is used in a 112-bit block. The block code has a minimum distance of five; there are patterns of three or more errors within the block for which the decoder output will be incorrect.[6] A 16-bit CRC as described in step 1 guards against an undetected *packet* (which may be many, many blocks) error.

Interleaving is used to reduce the susceptibility to burst errors. All 112 bits of every block are placed in a 7×16 matrix in column order, and read out in row order as shown in Figure 9-8.

	|<———————— column ————————>|										
row	**1**	**2**	**3**	**4**	**5**	**6**			**13**	**14**	**15**	**16**
1 |	D0	P3	D7	P10	D14	P17	D21	....	....	....	....	P52
2 |	P0	D4	P7	D11	P14	D18	....	....	....	....	P49	S5
3 |	D1	P4	D8	P11	D15	....	....	....	....	P46	S2	P53
4 |	P1	D5	P8	D12	....	....	....	....	P43	D47	P50	S6
5 |	D2	P5	D9	....	....	....	....	P40	D44	P47	S3	P54
6 |	P2	D6	....	....	....	....	P37	D41	P44	D48	P51	S7
7 |	D3	......	......	......	......	P34	D38	P41	D45	P48	S4	P55

Figure 9-8 *MDC4800 Interleaving Technique*

Detailed simulations[7] show that the decoder is sometimes overwhelmed by short burst errors whose placement is simply not optimum; that same error placement sometimes permits even longer bursts (~20 bits) to be corrected. Typically the interleaving approach permits correction of ~16-bit burst errors about 90 percent of the time; the CRC detects *most* of the balance. The published undetected bit error rate for MDC4800 is 1×10^{-5} (1 in 100,000).[8]

3. *Good error correction; interleaving; strong CRC backup:* This philosophy is employed in Motorola's latest analog protocol, RD-LAP. Obviously there are major technical improvements over MDC4800. FEC is achieved via a rate 3/4 Trellis Coded Modulation technique; interleaving is organized to permit correction of ~32-bit-burst errors; there are cascaded CRCs with a final CRC-32 to detect uncorrectable packet errors. A burst error as long as 32/19200 = ~1.7 milliseconds can always be safely handled, and very few longer bursts escape detection. The published undetected bit error rate is 1.4×10^{-11}, an astonishingly low 1.4 errors in 100 billion.
4. *Strong FEC; no interleave; no CRC:* The objective is to exploit the processing power of new "engines" so virtually all errors can be corrected without retransmission. The most common technique used is Reed-Solomon coding developed in 1960 and initially used in large file controllers capable of bearing the processing expense associated with this technique.

   By 1982 microprocessors permitted Reed-Solomon codes to be used in MDI's (now absorbed by Motorola) MMP-based vehicular terminals. The practical results were outstanding, with published UBER of 1 in 10 billion characters.[9]

Reed-Solomon is also planned for the CDPD protocol. This implementation is similar to the MDI approach, including the absence of a backup CRC (see Table 9-1), but CDPD has chosen to drive its code rate higher, possibly to match RD-LAP, to the

Table 9-1 *Reed-Solomon Comparison: CDPD versus MMP*

	CDPD	MMP
m (bits/symbol)	6	6
Symbol alphabet size (2^m)	2^6 (64)	2^6 (64)
k: encoder input symbols	47	45
: encoder input bits	282	270
n: $2^m - 1$ symbols	63	63
: $m(2^m - 1)$ bits	378	378
Code rate (k/n)	.75	.71
$n - k$: $2t$ symbols	16	18
: $2mt$ bits	96	108
t: symbol error-correcting ability	8	9
: bit error-correcting ability	48	54
Transmission speed (bps)	19,200	4,800
Fade duration protection (ms)	2.50	11.25

detriment of its error-correction capability. Note that the four-fold increase in CDPD transmission speed, combined with a higher code rate, results in ~20 percent of the fade protection achieved by MMP. The 2.5-millisecond fade protection is not the recommended implementation because it yields a poor undetected *block* error rate of 1.2×10^{-5} (1.2 in 100,000).[10] The recommended course is to correct only 7 symbols, resulting in an undetected block error rate of 2.75×10^{-8} (100 million).[11]

Note that CDPD does not employ CRC for burst error detection.

9.2.4 ARQ Alternatives

9.2.4.1 ARQ Variations

There are three principal ARQ variations:

1. *Stop-and-wait*, in which the sender transmits one variable-length packet and waits for an acknowledgment before the next packet is sent. If an ACK is received, the packet contents, which have been temporarily held in a buffer, are discarded and the next packet transmitted. If a NAK or time-out is received, the saved packet is retransmitted. This is the technique used by Motorola in MDC4800 and RD-LAP and is the simplest, lowest cost *device* implementation.
2. *Go-back-N*, in which multiple packets are sent continuously without waiting for acknowledgments. A ceiling is placed on the permissible number of blocks that can be outstanding without an ACK. If an error is detected, the sender must retransmit the error packet as well as all succeeding packets. This ensures that the blocks at the receiver are in the correct sequence with a minimum of processing but storage expansion is necessary. It results in modest device cost increases, but offers greater transmission efficiency for long messages. This is the technique used by Ericsson in Mobitex (RAM).
3. *Selective*, in which only the packet(s) in error are retransmitted. The device message buffer space is large, as it is in "go-back-N," but the device processing complexity is also increased as the packet sequence must be maintained. It offers superior transmission efficiency for long messages. This is the CDPD approach.

9.2.4.2 Practical Results

Although stop-and-wait approaches intrinsically "feel" inferior, this may not prove to be the case in real-world situations. For short, single-packet messages (such as ~100 octets) there is no practical difference between ARQ techniques. All must await the successful ACK indicator.

As message lengths rise, such as to 200–500 octets, Motorola (in RD-LAP) will transmit the message as a single packet, and Ericsson as a multiblock "go-back-N" packet. CDPD, however, will transmit this message as 2 to 5 individual packets. Long messages are more vulnerable to the occurrence of error sometime during their transmission. Thus, with an uncorrectable single burst error Motorola implementations will repeatedly try to resend a 500-octet message, exposing it to a new error each time; CDPD (the other extreme) will only retransmit the 100-octet packet containing the error, and its probability of successful receipt is improved.

As the number of errors in the message rise, the advantages of "go-back-N" and selective ARQ are blunted somewhat, as illustrated in Table 9-2.

Table 9-2 *Single Packet Failure in Message:*

Packets in Message	Average Packets Resent		
	Stop&Wait	Go-Back-N	Selective
1	1.0	1.0	1.0
2	—	1.5	1.0
3	—	2.0	1.0
4	—	2.5	1.0

Two Packet Failures in Message:

Packets in Message	Average Packets Resent		
	Stop&Wait	Go-Back-N	Selective
2	—	2.0	2.0
3	—	2.7	2.0
4	—	3.3	2.0

A more precise view of the value of selective ARQ can be seen in Figure 9-9. The CDPD V0.8 block success rate was modeled[12] in a burst error environment characterized by a C/N of 20 dB, at a speed of 30 mph (more on this later), with background Gaussian noise. The results were calculated on both a stop-and-wait and selective ARQ basis. With message lengths of 400–500 octets, overall throughput was improved by ~30 percent.

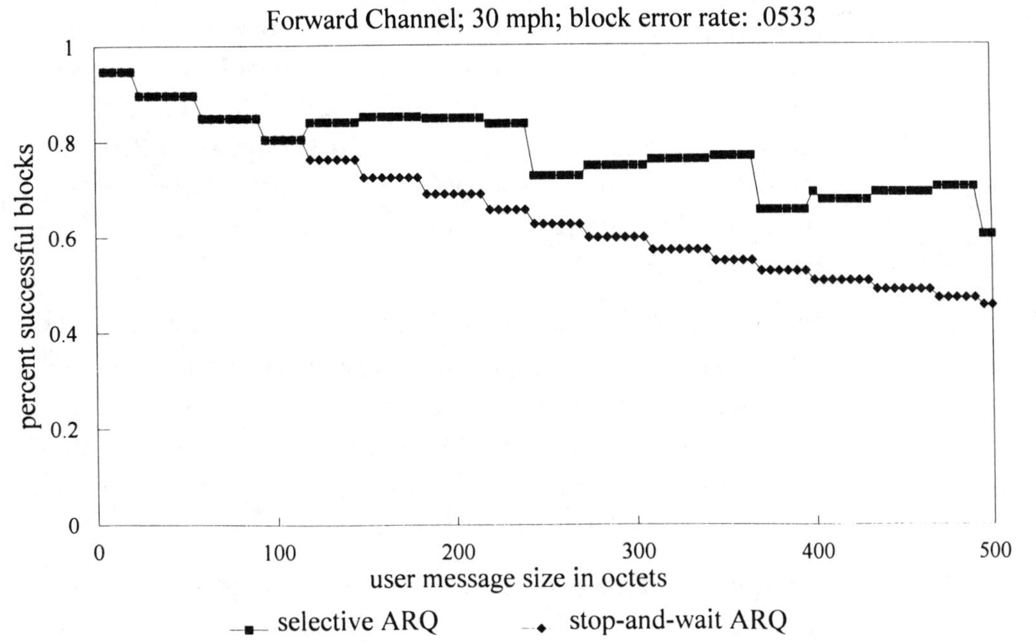

Figure 9-9 *CDPD V0.8 Block Success Rate*

9.2.5 A Data Flow Example

To illustrate the packet principles just covered, an outbound, multipacket message, with a single transmission error controlled by selective ARQ, is shown in Figure 9-10.

Base A Sends	**B, I S=6, R=4**	**B, I S=7, R=4 (error)**	**B, I S=0, R=4**	**B, I S=7, R=4**	**B, I, P S=1, R=4**	
Unit B Sends			**B, SREJ, F R = 7**			**B, RR, F R=2**

Figure 9-10 *Data Flow Example*

Using the Address/Control notations described in Section 9.1, Base **A** first sends an information *command* to **B** identifying this outbound packet as number 6; it also notifies **B** that the next packet expected from **B** is number 4. In the second step of the sequence **A** sends an information block numbered 7 that is damaged en route. In sequence 3 **A** sends another packet to **B** (note the 7-bit counter has wrapped from 7 to 0), unaware of the prior problem.

Meanwhile, **B** *responds* with its address and tells **A** that it is selectively rejecting packet number 7, which contained an error. The poll/final bit says that's all **B** has to say at this time. **A** sends **B** the corrected copy of packet 7, out of sequence, and then finishes the message by sending packet number 1. Perhaps a little nervous, **A** *polls* **B** for its status.

B *responds* with its address, says that everything is okay (it's in the receive ready state), that the next packet it expects to receive from **A** is number 2, and that that's all it has to say.

Simple, precise, unambiguous—and highly efficient.

9.3 FADE RATE VERSUS FADE DURATION

9.3.1 Characteristics of a Fading Channel

The robustness of the protocol's fade resistance varies with the speed of the target—vehicle or human. Before this can be explored, some background is required.

Typical signal levels due to fading cover attenuations ranging from a small increase in signal strength to decreases of -30 dB and beyond. The distribution of fade depths is statistical. At a given frequency, fade rate and duration depend upon both the fade depth and the velocity of the mobile radio. As velocity increases so does the fade rate. But the fade duration shortens as the radio moves in and out of the fading zone more quickly. The number of times per second that the signal crosses a specific dB level (the fade, or level crossing rate), and the duration of time it spends below that level (the fade duration), are variable. A fading signal crosses the -20 dB level, and spends less time below -20 dB, than it does at -10 dB.

9.3.2 Fade Rate

Figure 9-11 is a plot, with mobile unit velocity a parameter, of the fade rate at a variety of instantaneous signal-power levels relative to the mean value.

Figure 9-11 is derived from Reudink's[13] formula:

$$Nr = (2\pi)^{1/2} * F * A * e(-A^2)$$

where:

F = vehicle velocity ÷ radio wavelength

A = signal amplitude in the fade ÷ RMS signal level

Imagine a target moving at a constant *speed* and encountering a destructive fade. If the fade rate is more frequent than the packet rate, the protocol can never be free of the error. Each message *length* also has a critical fade rate, the point at which a destructive failure can never be shaken. For CDPD the smallest transmittable unit is an outbound block 420 bits (~25 user octets plus compressed addressing) long. For this particular length the critical fade rate is:

$$\frac{19{,}200 \text{ bps}}{420 \text{ bits}} = 45.71 \text{ Hz}$$

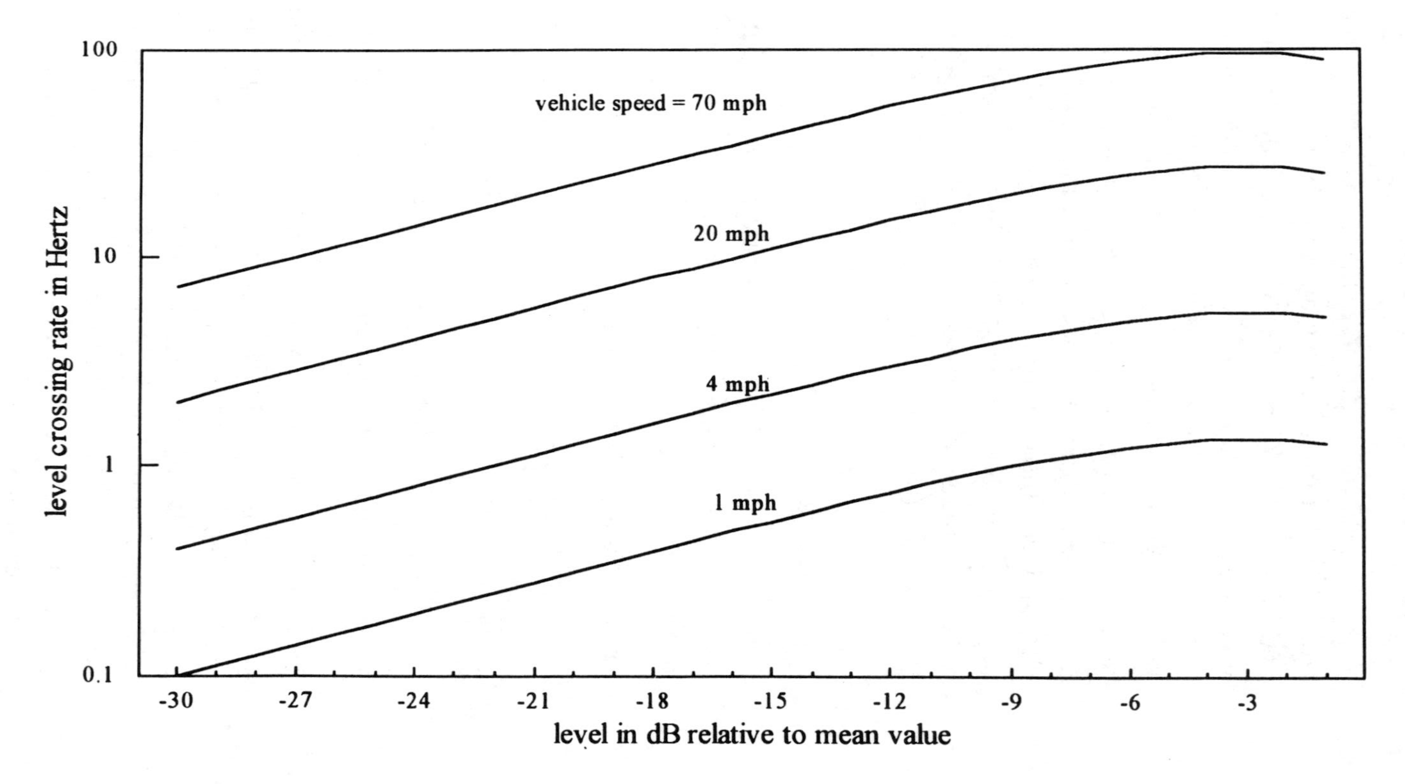

Figure 9-11 *Fade Rate versus Speed*

This critical rate intersects the 70 mph curve of Figure 9-11 at the "edge of coverage" (~17 dB). At this signal level and speed even a single-block message will continually fail. But if the target speed is reduced to 20 mph the critical fade rate is never reached. Velocity counts.

Virtually all protocols are length sensitive: The longer the message the more exposed to a damaging fade occurring within a message. If the CDPD packet contains 4 blocks (~126 user octets) the critical fade rate is 11.4 Hz and some misses will begin to occur even at 20 mph.

Consider two Motorola protocols, each with a packet containing 108 user octets, a realistic size for applications on ARDIS. The critical fade rate (Nr) for each is:

MDC4800[14]

$$\mathrm{Nr} = \frac{1.0}{\text{sync times} + \text{GIB/GRB/UCB times} + (n \times \text{block time})}$$

$$= \frac{1.0}{18.33 + 70.0 + (18 \times 23.33)} = \frac{1.0}{88.33 + 419.94} = \frac{1.0}{508.27}$$

$$= \mathbf{1.97\ Hz}$$

RD-LAP[15]

$$\mathrm{Nr} = \frac{1.0}{\text{preamble time} + (n \times \text{block time}) + \text{CRC time}}$$

$$= \frac{1.0}{12.19 + (9 \times 6.875) + 1.67} = \frac{1.0}{12.19 + 61.875 + 1.67} = \frac{1.0}{75.73}$$

$$= \mathbf{13.2\ Hz}$$

where n = number of encoded blocks

Clearly the low-speed (4800 bps), inefficient MDC4800 protocol, which needs a half-second to transmit 108 octets, will be exposed to fade *rate* problems. The much higher speed (19,200 bps) and more efficient RD-LAP protocol is much less sensitive to rate problems. A protocol exhibiting *rate* sensitivity is one in which the target velocity should be slow. High target speeds encounter more fades; thus MDC4800 (not surprisingly) has good rate characteristics for pedestrians using handheld devices.

9.3.3 Fade Duration

In like manner a plot of fade duration, the average time the fading signal spends below a given level, is shown in Figure 9-12.

Note that fade duration is inversely proportional to target speed: 70 mph is the bottom curve.

Like Figure 9-11 these curves are also derived from another Reudink formula:

$$t = \frac{e(A^2 - 1)}{(2\pi)^{1/2} * F * A}$$

Again, imagine a target moving at a constant speed and encountering a destructive fade. If that fade *duration* is longer than the correction mechanism can accommodate, the protocol can never be error free. It is of little help to always detect the failure; if it cannot be corrected the packet must fail. Length sensitivity is real. A long, multipacket message is more likely to see an unpleasant fade than a brief ACK.

Even short messages will be damaged at times. CDPD's recommended correction power is seven 6-bit symbols per block. Ignoring the effect of static bit errors (which are very damaging to Reed-Solomon correction techniques), a block can theoretically withstand a fade of 42 bits (2.2 milliseconds). Thus walk-speed targets will sometimes encounter fades of far greater duration than the error-correction code can possibly handle.

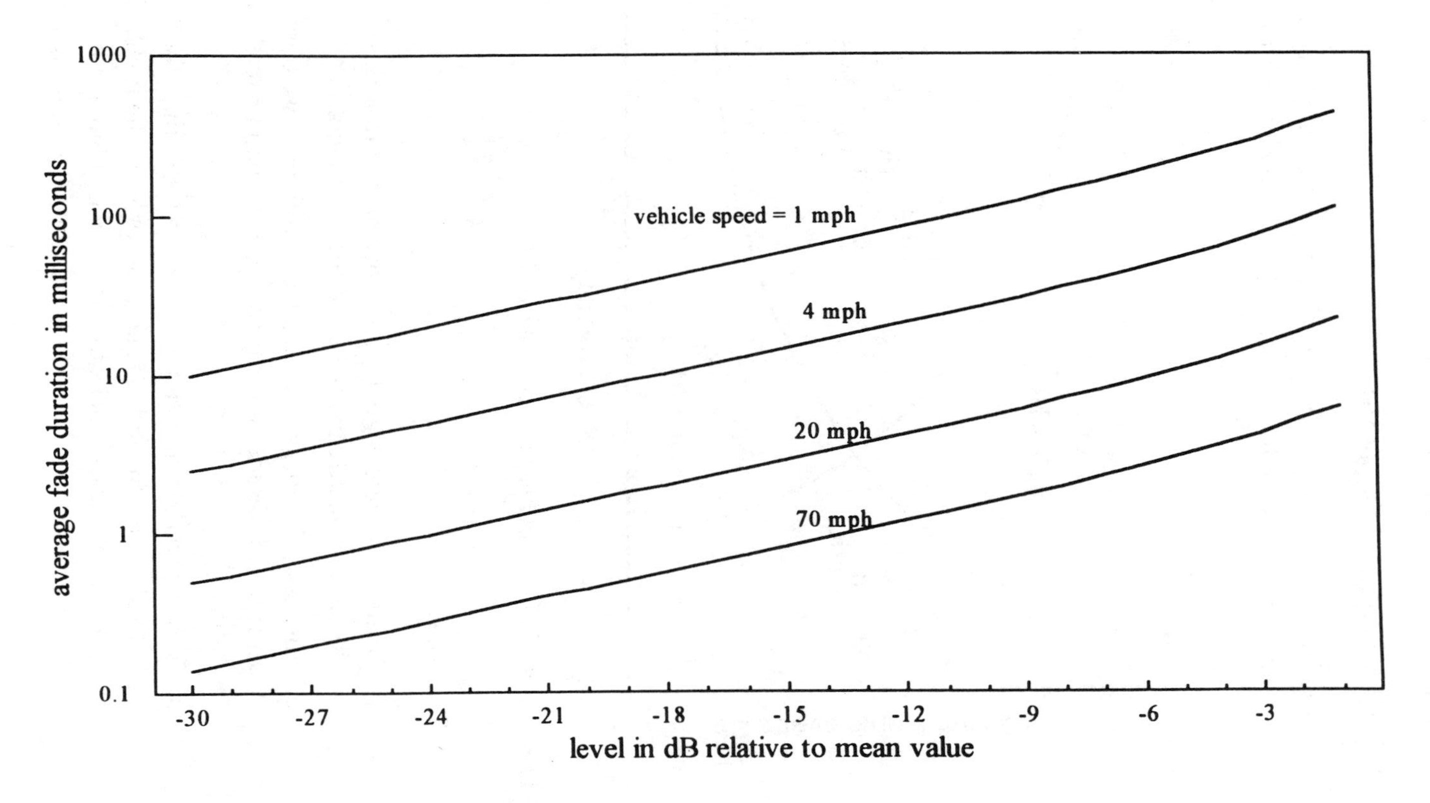

Figure 9-12 *Fade Duration versus Speed*

9.3.4 Optimum Target Velocity

Reudink's formulas can be graphically portrayed in an alternative manner to create four zones as shown in Figure 9-13.

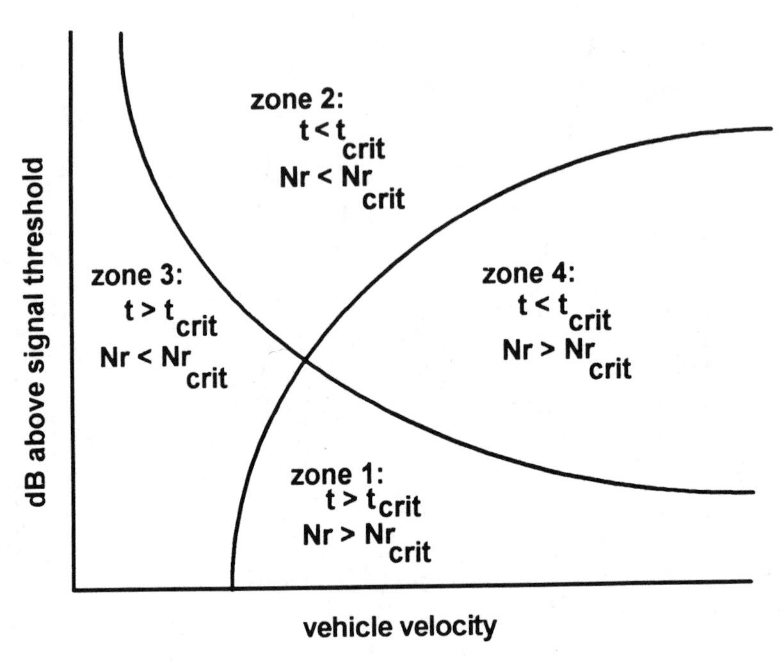

Figure 9-13 *Theoretical Throughput Curve by Velocity*

The zones are:

1. Failure: No blocks get through because both fade duration and fade rate exceed the critical thresholds.
2. Success: All blocks always get through because both fade duration and fade rate are always less than the critical values.
3. Retry: The fade duration is always greater than the critical value, but the fade rate is always less. The noncritical fade rate permits blocks to pass between fades. Retries will be necessary due to fade duration hits somewhere in

the block, but retransmission will eventually pass a successful block.

4. Retry: The fade duration is always less than the critical value, but the fade rate is greater. The frequent short fade bursts appear as random errors and cause retransmission when the cumulative effect is an uncorrectable block.

There are actually a class of curves that must be generated for each protocol, determined by:

1. The smallest size of the block that can be protected by the error correction/detection approach
2. The maximum fade duration that the FEC technique can correct

Even then no hard, fast boundaries between the four zones exists. The fade rate/fade duration equations are time-averaged values, and do not account for retransmission time-out effects. In practice, smooth transitions from one zone to another occur. Relative probabilities of successful transmission can be calculated that will produce 5 percent, 50 percent, and 95 percent curves that will resemble those shown in Figure 9-14. But note that these curves can never be precise. They do not account for random bit errors that overlay the fade, nor do they account for compensatory techniques such as antenna diversity that will exist in some devices.

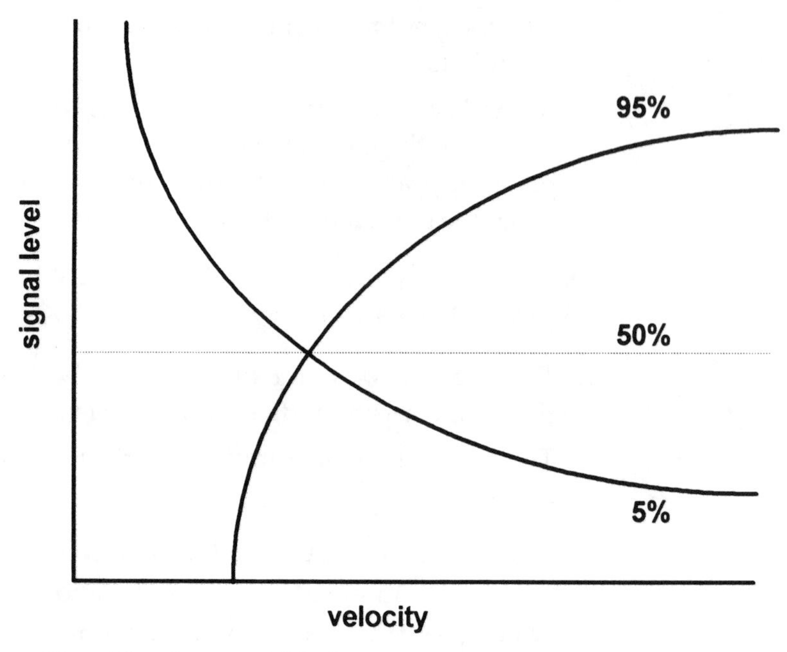

Figure 9-14 *Signal Level versus Velocity*

However, calculating the curves for each protocol gives a solid indication of its optimum design speed, the signal strength required, and the message-length sensitivity.

As an example, when the curves are displayed for the two 108-octet Motorola examples (Figure 9-15), an application design fit is suggested. MDC4800 will perform best at very slow target speeds, ideal for the field service person walking with device hanging from belt or shoulder. RD-LAP will be optimum on metro-speed vehicles. ARDIS, with frequency-hopping modems capable of operating on either protocol, can exploit this characteristic.

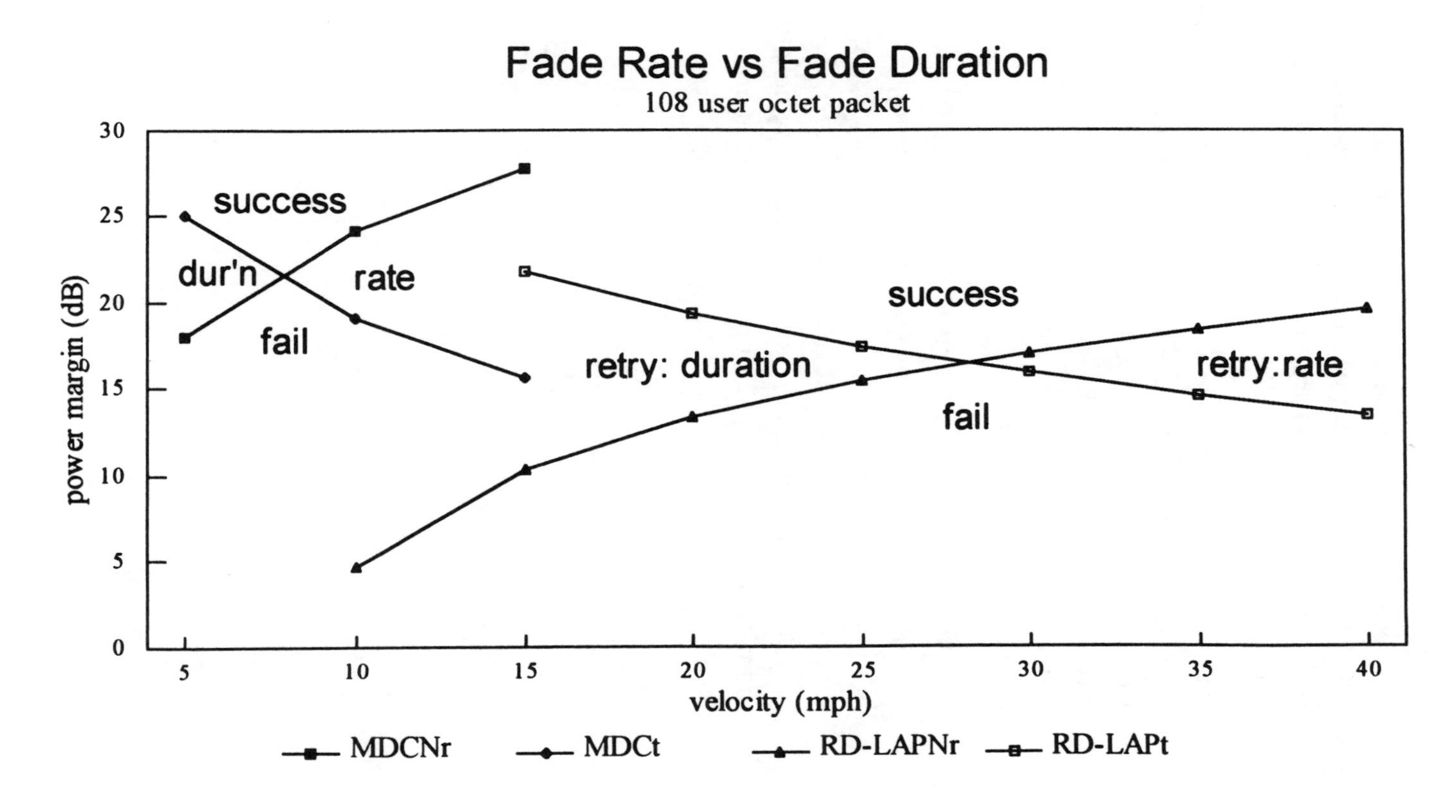

Figure 9-15 *Fade Rate versus Fade Duration: Motorola*

This does not mean that a single protocol will not operate at a suboptimum velocity. It does indicate that retransmission rates will likely rise—a capacity problem for the service provider, and a response-time problem for the user.

It is also possible to reduce retries by providing space diversity antennas in correctly engineered subscriber units. Motorola's KDT8xx family, and associated external modems, use dual-receive, switched-diversity antennas. This advantage can be lost if device vendors are unable, or choose not, to deploy diversity for product cost reasons.

The impact of message length can be illustrated by a CDPD example[16] (Figure 9-16). The curves for both single- and four-block messages indicate that very short CDPD messages can be successfully transmitted at relatively high target speeds with very low power margins. As the message lengthens the optimum speed falls below 25 mph. The power margin requirement also rises, but is still within cellular edge of coverage.

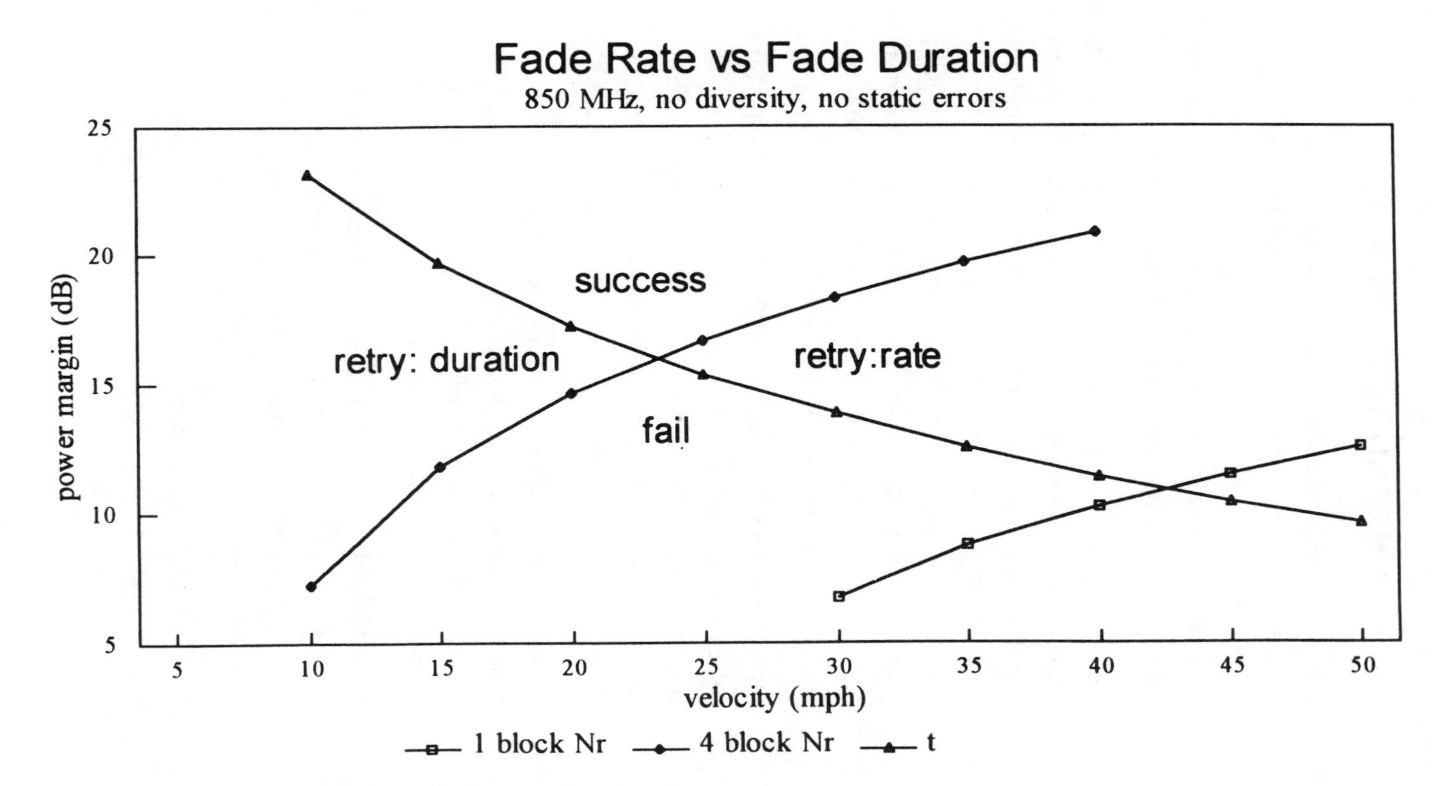

Figure 9-16 *Fade Rate versus Fade Duration: CDPD*

9.4 SYNCHRONIZATION ERRORS

An important contributor to message retry is the failure to achieve sync. In continuous transmission (usually outbound) the main problem tends to be frame sync; keyed transmission (usually inbound, but also outbound in implementations such as ARDIS) also has significant impact from bit sync.

The frame sync approach varies widely by vendor. Motorola's MDC4800 protocol uses a single-frame sync in the preamble of a packet. It is a pseudorandom pattern, 40 bits in length, with a correlation peak-to-side-lobe ratio of about 7 to 1. Many bit errors can be tolerated in the pattern. The incoming bits are slid past the sync pattern, multiplied bit by bit, and summed. A perfect answer is, of course, 40. But results below that number may still permit frame sync to be declared.

Simulation results[17] on MDC4800 show that if no bit errors are tolerated, failure to achieve frame sync will occur ~3 percent of the time. If 12 bit errors are permitted, false frame sync will be declared ~3 percent of the time. Each system can be tuned individually to eliminate most frame sync errors, missed or false.

Because the ARDIS implementation is keyed transmission, the optimum bit-sync length can also be simulated. At length 48, with a realistic error profile,[18] less than 1 percent of packets are retried because of bit-sync errors.

Note that Mobitex has a 16-bit pattern for both bit and frame sync. Only a single-bit error is tolerated in the frame-sync pattern. Sync error-message retries are significantly higher for Mobitex.

CDPD[19] V0.9 has a distributed 35-bit forward sync pattern, which does double duty as a busy/idle indicator, for each block. A decode pattern also can be viewed as part of the forward channel synchronization process. The error tolerances for these multiple syncs vary, but are generous; simulation after simulation[20] indicates that they are quite robust. Less than 1 percent fail with a representative error profile.

When bit-sync errors are included for the reverse channel, which also bears the brunt of false busy/idle and decode status indications, the overall "sync" failure rate approaches 2.5 percent with the simulated error profile.

9.5 INBOUND ACCESS: CONTENTION MODE

9.5.1 ALOHA

The simplest contention mechanism is ALOHA.[21] The throughput formula for ALOHA is:

$$S = Ge^{-2G}$$

where *S* (throughput) is the average number of *successful* transmissions per packet transmission time *P*, and *G* (offered traffic) is the average number of *attempted* packet transmissions per *P*.

This formula is a Poisson distribution in disguise, developed the following way:

Poisson distribution:

$$P\{k \text{ arrivals in } t \text{ seconds}\} = \frac{(At)^k}{k!} e^{-At}$$

where A = arrival rate in packets/second

If the average packet transmission time is length *P*, and colliding packets can begin any time, the vulnerable period is length 2*P*. A packet in process can have its very last bit ruined by an interloper (one *P* time), and the villain transmits through a second *P* even though its first bit has been destroyed.

It is also possible to define the relationship between *S* and *G* very simplistically:

$$S = GP\{\text{good transmission}\}$$

That is, throughput equals offered traffic times the probability that the transmission is good.

Define *P*{good transmission} as no additional transmissions during the vulnerable period 2*P*. Then in the Poisson distribution *k* becomes 0, as in *P*{0 arrivals in 2*P* seconds}. When $k = 0$ the fraction in the Poisson distribution vanishes. Further, *t* can be defined as 2*P*, and *A* (arrival rate) can be defined as *G*/*P*. This leaves:

$$P\ \{0 \text{ arrivals in } 2P\} = e^{-(G/P)*2P} = e^{-2G}$$

Now one can substitute this new value of *P* into the prior equation to obtain:

$$S = Ge^{-2G}$$

ALOHA contention schemes are still in use by some vendors (e.g., Electrocom Automation), but most designers reject the low maximum throughput (18.4 percent) and instability of this technique. The maximum throughput can be achieved (in an ideal world) when the offered traffic reaches 50 percent. Drive the channel harder and the throughput actually drops, leading to chaos. The two most common contention alternatives follow.

9.5.2 Slotted ALOHA

This simple refinement of ALOHA, developed initially for satellite channels, forces packets to begin only on defined boundaries. Using the ALOHA example these slots are exactly one *P*. This halving of the vulnerable period converts the ALOHA formula to:

$$S = Ge^{-G}$$

and doubles the throughput potential. Now the inbound channel can theoretically reach a useful throughput of 36.8 percent when the channel is 100 percent loaded. There is more overhead to consider, and the channel can still be unstable (but manageable); still, the throughput draws vendors.

9.5.3 Slotted CSMA

Of all the myriad varieties of carrier sense protocols, the most common candidate is slotted CSMA. This technique is used in RD-LAP, Mobitex, and CDPD. As always, each vendor has its own variation but the principle is the same: Sense the outbound channel for information about the status of the inbound channel; if the channel appears free, transmit as slotted ALOHA; if the channel is busy, back off a random interval; in either case when the receiver detects inbound traffic, place channel-busy information in the outbound channel for the next round of sensing.

The equation for slotted nonpersistent CSMA is:

$$S = \frac{aGe^{-aG}}{1 - e^{-aG} + a}$$

with *a*, the ratio of time without knowledge of the channel state to the average transmission time. This is a new variable often called the collision window. The goal is to keep it as short as possible. If the average transmission time is 100 milliseconds, and the collision window is also 100 milliseconds, CSMA is worse than useless. Two devices could be ready, each sense the channel free, destructively collide upon transmission, and complete their transmissions, and then the channel is signaled busy for no good purpose. This variable *a* determines the throughput of slotted CSMA.

Half-duplex systems such as Motorola's have an inherent disadvantage when designing collision windows. When a packet is ready the device "listens" for an outbound signal indicating channel status. Sensing channel free, it switches off the receiver to reprogram the frequency synthesizer. At this instant the inbound channel is busy because the decision to transmit is irrevocable. When the synthesizer locks on to the desired frequency the transmitter is keyed. It rises to full power, often simultaneously transmitting sync patterns, until message transmission begins. This interval is driven by transceiver characteristics. In an excellent system the time period will be on the order of 5 milliseconds.

Only now can the base station act on this signal, even though damage may already have occurred to a simultaneous device's transmission. The busy channel status is set 3–4 milliseconds later, if appropriate (the count field may indicate otherwise), and the devices must sense repetitive busy indicators to avoid falsing. It is not uncommon to find collision window intervals of 10–15 milliseconds in half-duplex systems.

The collision window, and the average packet length, constrain the achievable CSMA throughput. For RD-LAP an error-free *S* of ~40 percent is probably achievable on realistic message lengths with a *G* of 80 percent.[22] But note that this means that each message, on average, tries twice before succeeding—and this is before retransmission due to errors.

The CDPD base design is full duplex because the data devices must always listen for voice in order to scramble out of the way. CDPD exploits this reality by shortening its collision window dramatically. Further, extraordinary efforts are made to shorten the busy hang time—the interval in which the channel is actually free but the devices are not yet aware of it. CDPD can signal busy within one slot time (60 bits): 3.13 milliseconds. The busy hang time varies with message length, and even the nature of vendor devices, but has the potential to be very short indeed under certain circumstances. CDPD appears capable of achieving an error-free *S* of 40 percent when the channel is loaded only to 70 percent.

9.6 RETRANSMISSION RATES

It is clear that the actual number of successful transmissions, especially inbound, is a fraction of those attempted. Contention constraints halve the attempts, and those packets that escape this trial are beset by new torments. Depending upon message length, target speed, and the ambient noise/fade conditions, the successful messages are slashed still further and must be repeated.

An examination of the MDC4800 protocol used in ARDIS can be instructive.[23] Figure 9-17 portrays the message success rate when burst errors occur, on average, at 2560-bit intervals with fixed lengths of 16, 18, and 20 bits. Background Gaussian noise is also present; a bit is marked to fail, on average, every 430 bit times.[24] Under these conditions, with 18-bit burst errors, ~95 percent of 20-byte messages would succeed on the first attempt, but only ~60 percent of the 240-byte messages. Message-length sensitivity is obvious.

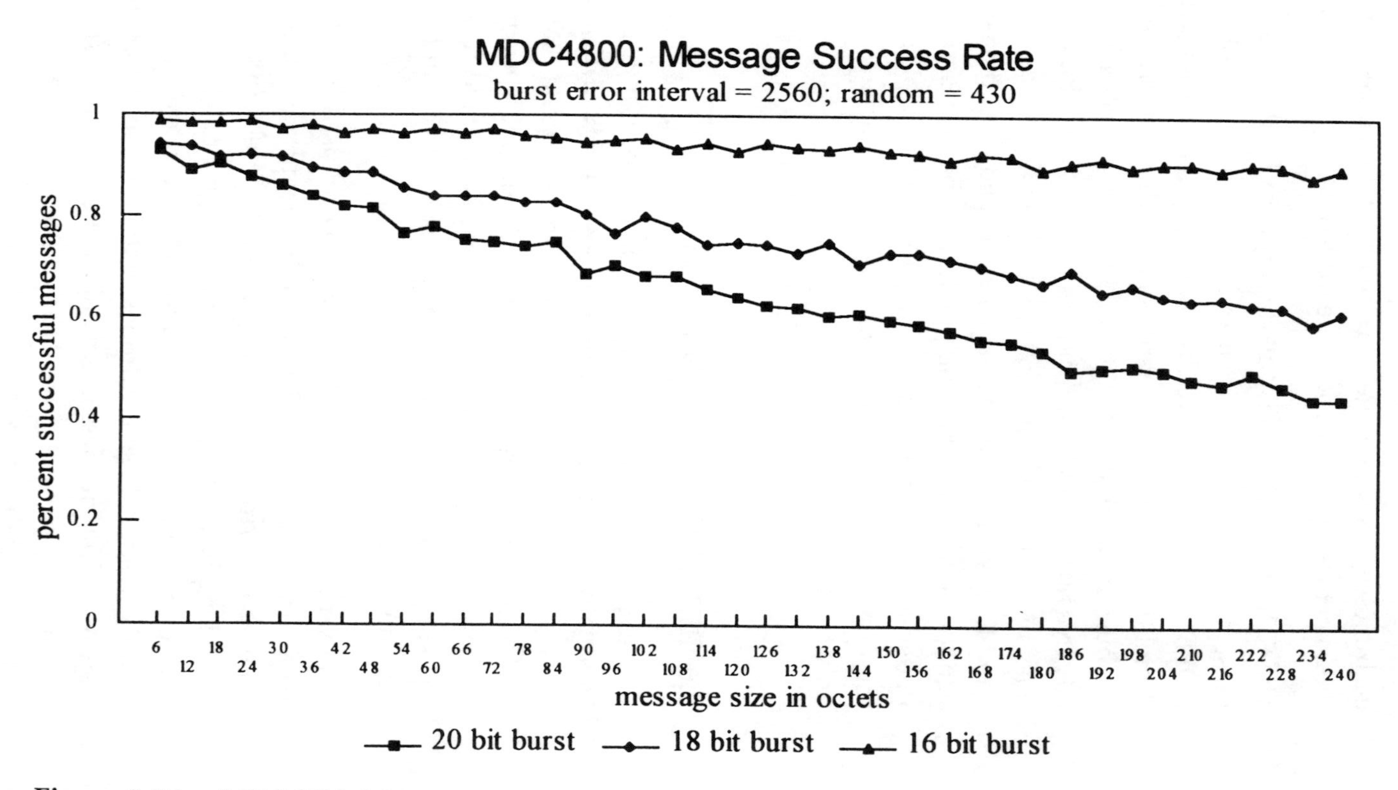

Figure 9-17 *MDC4800: Message Success Rate*

IBM has reported[25] that DCS (now ARDIS) experienced 15 percent retry rates with message sizes ranging from 62 to 108 octets. ARDIS itself now reports[26] a 90 percent success rate on the initial transmission of ~110-octet messages, 6.2 percent on the first retry, 2.8 percent on the second retry, with only 1 percent going to a third retry.

This suggests that a real-world fade duration of 18 bits (3.75 ms) is too high at 4800 bps. The actual is probably closer to 17 bits (3.5 ms), and that number was used for subsequent estimates. Note the implication of this fade duration interval on higher speed protocols such as RD-LAP and CDPD.

Table 9-3 illustrates that with short messages (up to ~60 octets) no more than two retransmissions are required to achieve message success. If the message exceeds 200 octets not only is a fourth retry occasionally required, but also a very small number of messages, 1 in 1,000, are not successful.

Table 9-3 *Successful Message Transmission Rate*

	%Messages Successful on					
	1st xmission	**1st retry**	**2nd retry**	**3rd retry**	**4th retry**	**Percent success**
6	0.980	0.020	0.000	0.000	0.000	1.000
12	0.974	0.026	0.001	0.000	0.000	1.000
18	0.968	0.031	0.001	0.000	0.000	1.000
24	0.961	0.037	0.001	0.000	0.000	1.000
30	0.955	0.043	0.002	0.000	0.000	1.000
36	0.949	0.048	0.002	0.000	0.000	1.000
42	0.943	0.054	0.003	0.000	0.000	1.000
48	0.937	0.059	0.004	0.000	0.000	1.000
54	0.930	0.065	0.005	0.000	0.000	1.000
60	0.924	0.070	0.005	0.000	0.000	1.000
66	0.918	0.075	0.006	0.001	0.000	1.000
72	0.912	0.080	0.007	0.001	0.000	1.000
78	0.906	0.085	0.008	0.001	0.000	1.000
84	0.899	0.090	0.009	0.001	0.000	1.000
90	0.893	0.095	0.010	0.001	0.000	1.000
96	0.887	0.100	0.011	0.001	0.000	1.000
102	0.881	0.105	0.013	0.001	0.000	1.000
108	0.875	0.110	0.014	0.002	0.000	1.000
114	0.868	0.114	0.015	0.002	0.000	1.000
120	0.862	0.119	0.016	0.002	0.000	1.000
126	0.856	0.123	0.018	0.003	0.000	1.000
132	0.850	0.128	0.019	0.003	0.000	1.000
138	0.844	0.132	0.021	0.003	0.001	1.000
144	0.837	0.136	0.022	0.004	0.001	1.000

Table 9-3 *Successful Message Transmission Rate (continued)*

	%Messages Successful on					
	1st xmission	1st retry	2nd retry	3rd retry	4th retry	Percent success
150	0.831	0.140	0.024	0.004	0.001	1.000
156	0.825	0.144	0.025	0.004	0.001	1.000
162	0.819	0.148	0.027	0.005	0.001	1.000
168	0.813	0.152	0.029	0.005	0.001	1.000
174	0.806	0.156	0.030	0.006	0.001	1.000
180	0.800	0.160	0.032	0.006	0.001	1.000
186	0.794	0.164	0.034	0.007	0.001	1.000
192	0.788	0.167	0.035	0.008	0.002	1.000
198	0.782	0.171	0.037	0.008	0.002	1.000
204	0.775	0.174	0.039	0.009	0.002	0.999
210	0.769	0.178	0.041	0.009	0.002	0.999
216	0.763	0.181	0.043	0.010	0.002	0.999
222	0.757	0.184	0.045	0.011	0.003	0.999
228	0.751	0.187	0.047	0.012	0.003	0.999
234	0.744	0.190	0.049	0.012	0.003	0.999
240	0.738	0.193	0.051	0.013	0.003	0.999

Table 9-4 demonstrates this phenomonon another way. If 100 6-octet packets are to be sent, the system will have to make 102 transmissions to successfully transmit them. But it will have to make more than 135 transmissions to successfully send the same number of 240-octet packets.

Table 9-4 *Retries Required for Successful Transmission*

			Successive Attempts				
	Retry rate	Initial attempt	1st retry	2nd retry	3rd retry	4th retry	Total xmission
6	0.980	1	0.020	0.000	0.000	0.000	1.000
12	0.974	1	0.026	0.001	0.000	0.000	1.027
18	0.968	1	0.032	0.001	0.000	0.000	1.033
24	0.961	1	0.039	0.001	0.000	0.000	1.040
30	0.955	1	0.045	0.002	0.000	0.000	1.047
36	0.949	1	0.051	0.003	0.000	0.000	1.054
42	0.943	1	0.057	0.003	0.000	0.000	1.061
48	0.937	1	0.063	0.004	0.000	0.000	1.068
54	0.930	1	0.070	0.005	0.000	0.000	1.075
60	0.924	1	0.076	0.006	0.000	0.000	1.082
66	0.918	1	0.082	0.007	0.001	0.000	1.089
72	0.912	1	0.088	0.008	0.001	0.000	1.097
78	0.906	1	0.094	0.009	0.001	0.000	1.104
84	0.899	1	0.101	0.010	0.001	0.000	1.112
90	0.893	1	0.107	0.011	0.001	0.000	1.120
96	0.887	1	0.113	0.013	0.001	0.000	1.127
102	0.881	1	0.119	0.014	0.002	0.000	1.135
108	0.875	1	0.125	0.016	0.002	0.000	1.143
114	0.868	1	0.132	0.017	0.002	0.000	1.151
120	0.862	1	0.138	0.019	0.003	0.000	1.160
126	0.856	1	0.144	0.021	0.003	0.000	1.168
132	0.850	1	0.150	0.023	0.003	0.001	1.177

Table 9-4 *Retries Required for Successful Transmission (continued)*

			Successive Attempts				
	Retry rate	**Initial attempt**	**1st retry**	**2nd retry**	**3rd retry**	**4th retry**	**Total xmission**
138	0.844	1	0.156	0.024	0.004	0.001	1.185
144	0.837	1	0.163	0.026	0.004	0.001	1.194
150	0.831	1	0.169	0.028	0.005	0.001	1.203
156	0.825	1	0.175	0.031	0.005	0.001	1.212
162	0.819	1	0.181	0.033	0.006	0.001	1.221
168	0.813	1	0.187	0.035	0.007	0.001	1.230
174	0.806	1	0.194	0.037	0.007	0.001	1.240
180	0.800	1	0.200	0.040	0.008	0.002	1.249
186	0.794	1	0.206	0.042	0.009	0.002	1.259
192	0.788	1	0.212	0.045	0.010	0.002	1.269
198	0.782	1	0.218	0.048	0.010	0.002	1.279
204	0.775	1	0.225	0.050	0.011	0.003	1.289
210	0.769	1	0.231	0.053	0.012	0.003	1.299
216	0.763	1	0.237	0.056	0.013	0.003	1.310
222	0.757	1	0.243	0.059	0.014	0.003	1.320
228	0.751	1	0.249	0.062	0.016	0.004	1.331
234	0.744	1	0.256	0.065	0.017	0.004	1.342
240	0.738	1	0.262	0.069	0.018	0.005	1.353

Note that these losses occur *after* contention impact on the inbound side. But do not think the task impossible; ARDIS guarantees response times (a crude indicator of retry activity) in its contracts, a strong indicator that the obstacles can be overcome.

REFERENCES

[1] *CDPDP V0.8 System Specification*, Book 3, Figure 403-5, p. 403–18.
[2] Uyless Black, *Computer Networks: Protocols, Standards, and Interfaces*, p. 91.
[3] Racal-Vodata Publication No. VT1/1005/1085/L0-1.
[4] Voice and Data Transmission, G. A. Arredondo et al., *Bell System Technical Journal*, January 1979.
[5] J. L. Hammond and P. J. P. O'Reilly, *Performance Analysis of Local Computer Networks*, p. 65.
[6] JFD Associates simulation: decoder.pas.
[7] JFD Associates simulation: motorola.pas.
[8] Motorola specifications on 9100-11 Mobile Data Terminal.
[9] MDI 9031 promotional literature.
[10] *CDPD V0.8 System Specification*, Book 3, Section 6.1.1, footnote 7.
[11] Ibid.
[12] JFD Associates, CDPD V0.8 Airlink Capacity Analysis.
[13] "Properties of Mobile Radio Propagation above 400 MHz," *IEEE Transactions on Vehicular Technology*, vol. VT-23, November 1974, Douglas O. Ruedink, pp. 143–159.
[14] JFD Associates, MDC4800 Performance Analysis.
[15] From the March 30, 1992, "As Is" public protocol document.
[16] JFD Associates, CDPD V0.8 Airlink Capacity Analysis.
[17] JFD Associates simulation, motosync.pas.
[18] "Experimental Evaluation of Packet Error Rates for the 850 MHz Mobile Channel," *IEEE Proceedings*, August 1985, H. M. Hafez et al., p. 373.
[19] *CDPD V0.9 System Specification*, Book 3, Section 4.5, p. 402–19.
[20] JFD Associates, CDPDP V0.8 Airlink Capacity Analysis.
[21] N. Abramson, *The ALOHA System*, Computer Communication Networks.
[22] JFD Associates, RD-LAP Performance Analysis, Pass 3.
[23] JFD Associates, MDC4800 Capacity.
[24] JFD Associates: simulation model mototest.pas.
[25] Michael Boyt, Minutes of Mobile Data Conference, as reported in *Mobile Data Report*, 6-19-89.
[26] Tom Berger, ARDIS Vice-President of Radio Network & Product Technology, Lexington Conference, 7-27-93.

CHAPTER

10

COVERAGE VERSUS CAPACITY

10.1 ALTERNATIVE APPROACHES

Key tradeoffs must always be made between capacity and coverage. In many metropolitan areas RAM and ARDIS deploy about the same number of base stations. Their *geographic* coverage in those metro areas is similar. But their different technical philosophies yield very different practical results:

1. RAM, with its multiple-channel, cellular-reuse approach has the potential for high user capacity at the cost of some reduced in-building penetration.
2. ARDIS, with its single-channel, contiguous-layer approach concentrates on in-building coverage at the cost of lowered capacity potential.

The differences flow from fundamental decisions on channels allocated for data versus control, reuse philosophy, message redundancy, base-station location, bit rates, and transmit power levels.

CDPD will not be exactly comparable to the existing packet switched systems because, in a full-blown implementation, it would deploy *many* more base stations. Some carriers are hesitant about the scope of the deployment. Bell Atlantic Mobile states: "CDPD will be offered in markets where it is expected to be commercially viable, although equipment will not be installed at all cell sites in those markets."[1] This could mean far wider geographic coverage in time; in the short term is likely to mean just metro area coverage. If full metro area rollout is provided, CDPD's infrastructure should permit very high capacity, but does not guarantee excellent in-building penetration. A discussion of the systems follows.

10.2 RAM

RAM configures its base stations for maximum user capacity. Note that the Mobitex traffic models assume relatively short data messages with an average length of 50 octets.[2] The response time criteria for these messages is not demanding: "average . . . time for status messages less than three seconds, 95 percent less than 20 seconds; for data/text . . . add the time for the length of the message, max 6 seconds."[3]

With these assumptions "a data channel at a single base radio station can cope with data traffic for 1000–1500 mobile terminals . . . . Since the traffic channels are frequency planned so that interference should not occur, the data traffic on the traffic channels can be run . . . independently of the traffic at the other base radio stations."[4] This represents an enormous traffic carrying capacity, enhanced by the cellular-like frequency reuse plan.

In major metropolitan areas such as New York City or Los Angeles, RAM has 30 frequency pairs allocated, each projected to have a capacity of 1,000 short-message users. RAM expects to be "able to reuse them easily four times" for a single metro area capacity of 120,000 users.[5]

Unlike the ARDIS reuse technique previously depicted in Figure 7-1, RAM uses a cellular-like approach. Assume a simple 7-channel reuse plan as shown in Figure 10-1.

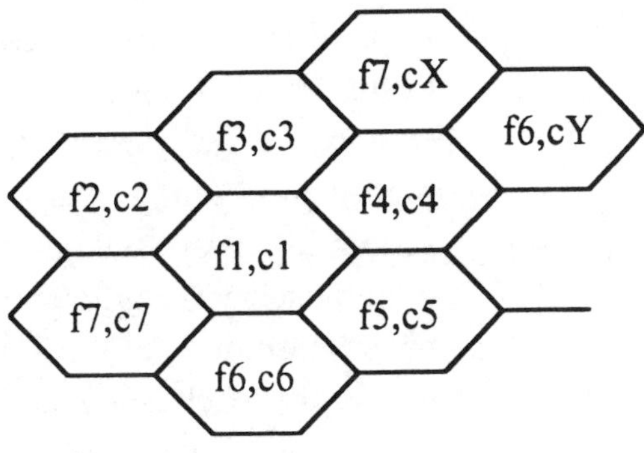

Figure 10-1 *Mobitex Channel Example*

If two (or more) subscriber unit messages collide inbound on frequency 7, in cell 7, retries will continue on this frequency/cell combination until a unit moves out of range and registers to a different base. The inbound messages will not be heard on frequency 7, cell X, because, by design, the distance between those two base stations is $\sqrt{3N}$ times the major radius of the cells. The Mobitex approach was initially vehicle oriented. Even the initial devices were high (6 watt) transmit power units.[6] Thus movement away from a troublesome base station was a very reasonable assumption. This philosophy began to change with the arrival of the low-power, handheld Mobidem.

Reports of penetration capability are currently anecdotal, though RAM is known to be taking extensive in-building measurements. It should be somewhat less thorough than cellular.

10.3 ARDIS

10.3.1 Coverage Philosophy

ARDIS configures its base stations for maximum in-building penetration and contractually guarantees its 90 percent coverage contours. An inbound message is received by multiple base stations, all tuned to the same frequency. The "typical" user transmission in Chicago is heard by 8 to 10 base stations.[7] Collision sensitivity is reduced and deep penetration coverage improved by this adjacent cell frequency diversity approach.

Recall that if two messages collide in cell 7, message "a" may be heard by cell 2; message "b" may be heard by cell 6. The base stations dutifully pass all clean copies of the messages to a higher processing level. Here alternative paths to each subscriber unit may well be simultaneously invoked: Cells 2 and 6 would each carry unique messages on the same frequency. Signal strength indicators and even FM capture play pivotal roles in this approach (see Figure 10-2). Further, if there are simultaneous messages on cell 7 and cell 4, the cellular distance rule causes them to be heard independently.

ARDIS Example

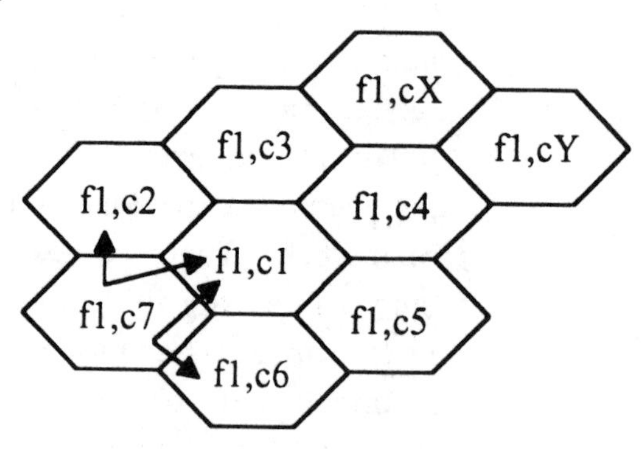

Figure 10-2 *ARDIS Example*

But all this keying and blocking of transmissions causes dead time on outbound base stations. Thus, for the same number of "cells," ARDIS does not have the subscriber capacity potential of RAM. The *national* capacity reported[8] at YE91 was 800,000 subscribers; the coming of 19,200 bps capability has probably pushed that potential over the 1 million mark. But it is clear that ARDIS deliberately sacrifices capacity in order to achieve the deep penetration coverage that forms the basis of its marketing strategy.

10.3.2 ARDIS Coverage versus Cellular

Given the important difference in total base stations, ~1,400 ARDIS versus ~10,000 all cellular, it is easy to assume that ARDIS coverage is only fractionally equivalent to that of cellular carriers. As usual, simplicity is often inaccurate. It is fair to say that cellular covers large tracts of geography where ARDIS (and RAM even more so) has no base stations at all. Example:

California counties	58
CA counties with at least one cellular base station	56
CA counties with at least one ARDIS base station	25

Thus, a potential ARDIS user with a need for coverage in Alpine, Amador, Butte, Calaveras, Colusa, Del Norte, Eldorado, and other remote California counties is simply out of luck unless the application fits the planned[9] ARDIS "complementary networks": American Mobile Satellite, Coverage-PLUS, and so on. Fortunately for ARDIS the current business population is concentrated in urban areas. But there will be customers lost because of lack of geographic span.

Where ARDIS chooses to compete, base-station siting is vastly different than cellular. The emphasis is on deep penetration achieved with relatively few, overlapping, high-power, same-frequency base stations.

An example of these differences can be demonstrated by inspecting San Diego County. ARDIS has a total of eight 45-watt base sites using frequency 856.65 MHz,[10] as shown in Table 10-1.

Table 10-1 *ARDIS Base Stations: San Diego*

	Address	ERP	Elevation	Latitude	Longitude
1	Mount San Miguel	71	2,565	32-41-47	116-56-06
2	530 B Street	53	70	32-43-05	117-09-32
3	3111 Camino Del Rio N	130	40	32-46-34	117-07-30
4	San Marcos Mountain	115	1,510	33-12-53	117-11-15
5	Mount Woodson	130	2,894	33-00-34	116-58-11
6	9750 Miramar Road	130	506	32-53-40	117-07-05
7	5252 Balboa Avenue	950	1,365	32-49-10	117-10-55
8	1200 Harbor Drive N	132	15	33-13-03	117-23-48

For illustration purposes only, without being constrained by real-world considerations (propagation over water, foliage loss, obstructions), these base stations can be represented on an area map as omnidirectional circles. Base stations 4 and 8 can be ignored because they serve the northwestern portion of the county, well away from San Diego proper. Base stations 1 and 5 can be temporarily set aside because they are aimed at the eastern third of the county.

The "core" coverage comes from just four overlapping base stations whose centers run on a north-south line from Miramar to the airport. Assume that base stations 2 and 3 have a 7.5-mile radius and that base stations 6 and 7 have an 8-mile radius. These are conservative assumptions given the height and power of the radios, but are adequate for this demonstration. A first cut coverage map of San Diego is shown in Figure 10-3.

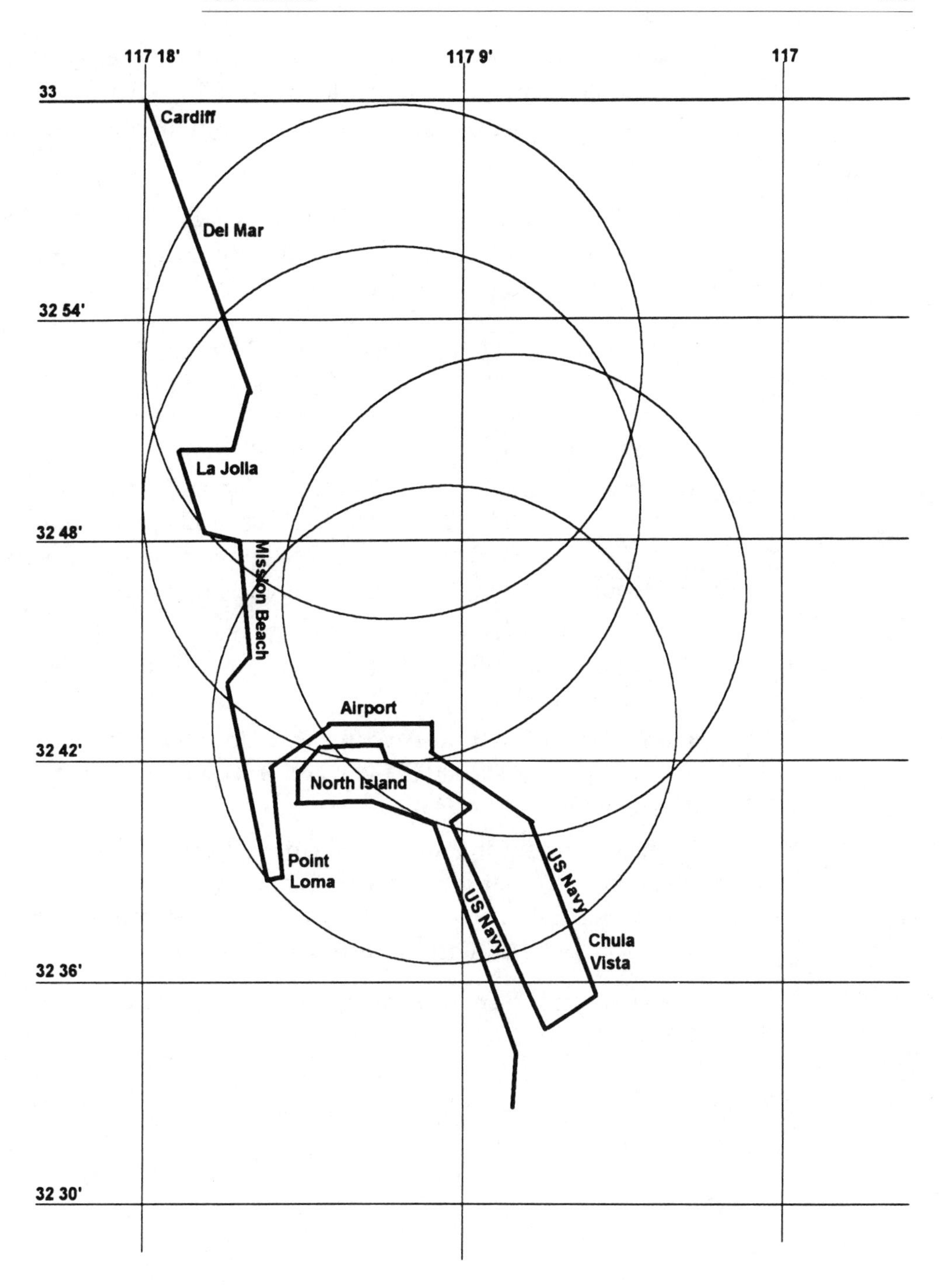

Figure 10-3 *ARDIS Core Coverage: San Diego*

For completeness, base stations 1 and 5 can be displayed to demonstrate their "gap-filler" role. Because of either the high ERP or elevation, a 10-mile radius was chosen for these circles, shown in Figure 10-4.

Although this ARDIS coverage example is only illustrative, the overall philosophy depicted is directionally correct. There will be:

1. Uncovered zones. The circles shown here, which are centered accurately, indicate that Cardiff to Solana Beach in the northwest and Imperial Beach to Chula Vista in the southwest may have poor (or no) coverage.
2. Areas illuminated by a single base station. Focusing on the "core," and ignoring the La Mesa or El Cajon areas, indications are that the Carmel Valley and Mission Gorge Roads (for example) have only single coverage.
3. Relatively rich areas illuminated by two base stations. Examples include the La Jolla/University City/Pacific Beach/Mission Bay area.
4. A large area of triple coverage including Miramar Naval Air Station, Lindbergh Field, and Balboa Park (Naval Hospital).
5. A critical core area with quadruple coverage. This sector is centered near the intersection of I-805 and Route 163 and includes key portions of I-8 and I-15 as well—a testimony to the California lifestyle.

A visual illustration of these areas of overlap is shown in Figure 10-5.

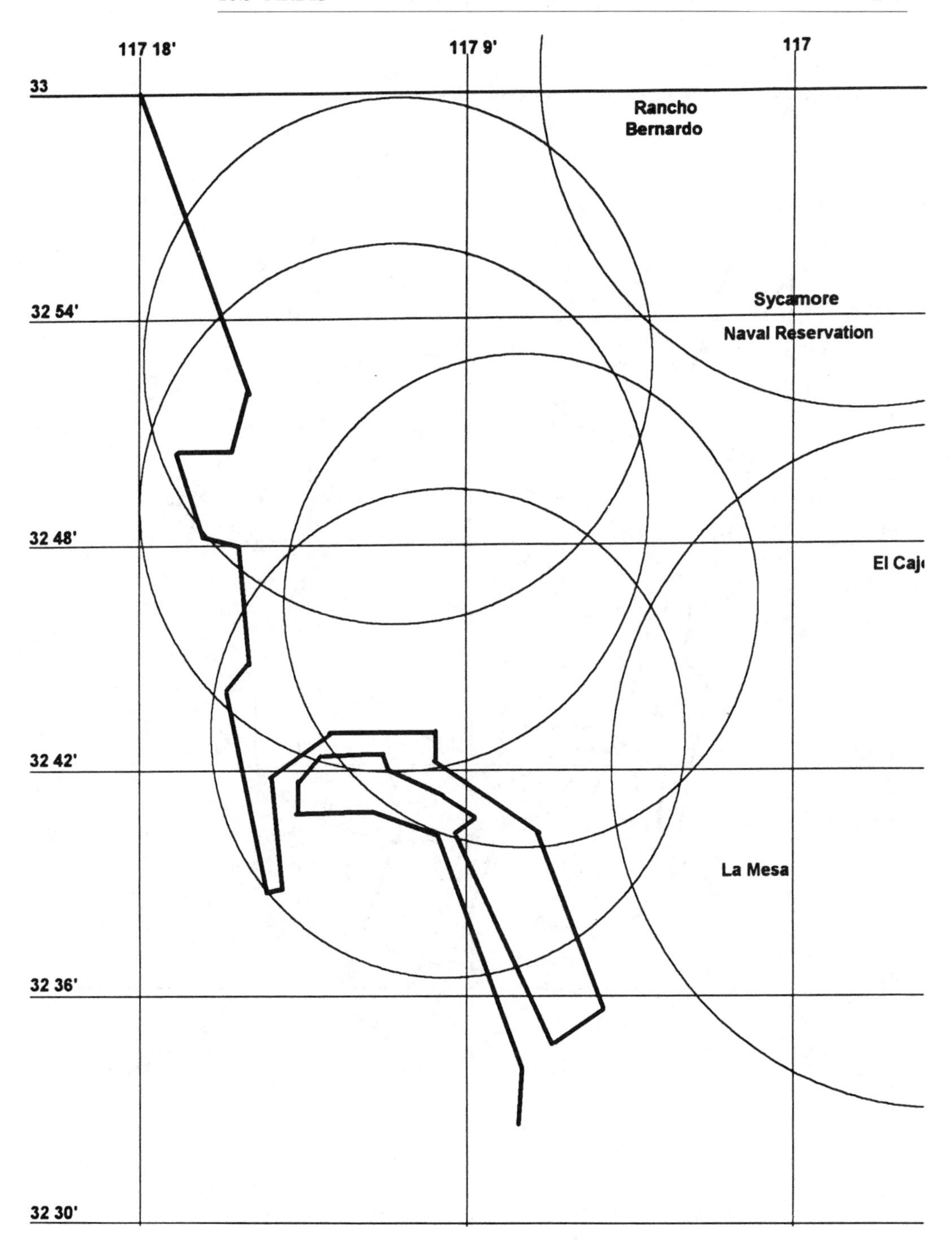

Figure 10-4 *ARDIS Overlap Coverage Zones: San Diego*

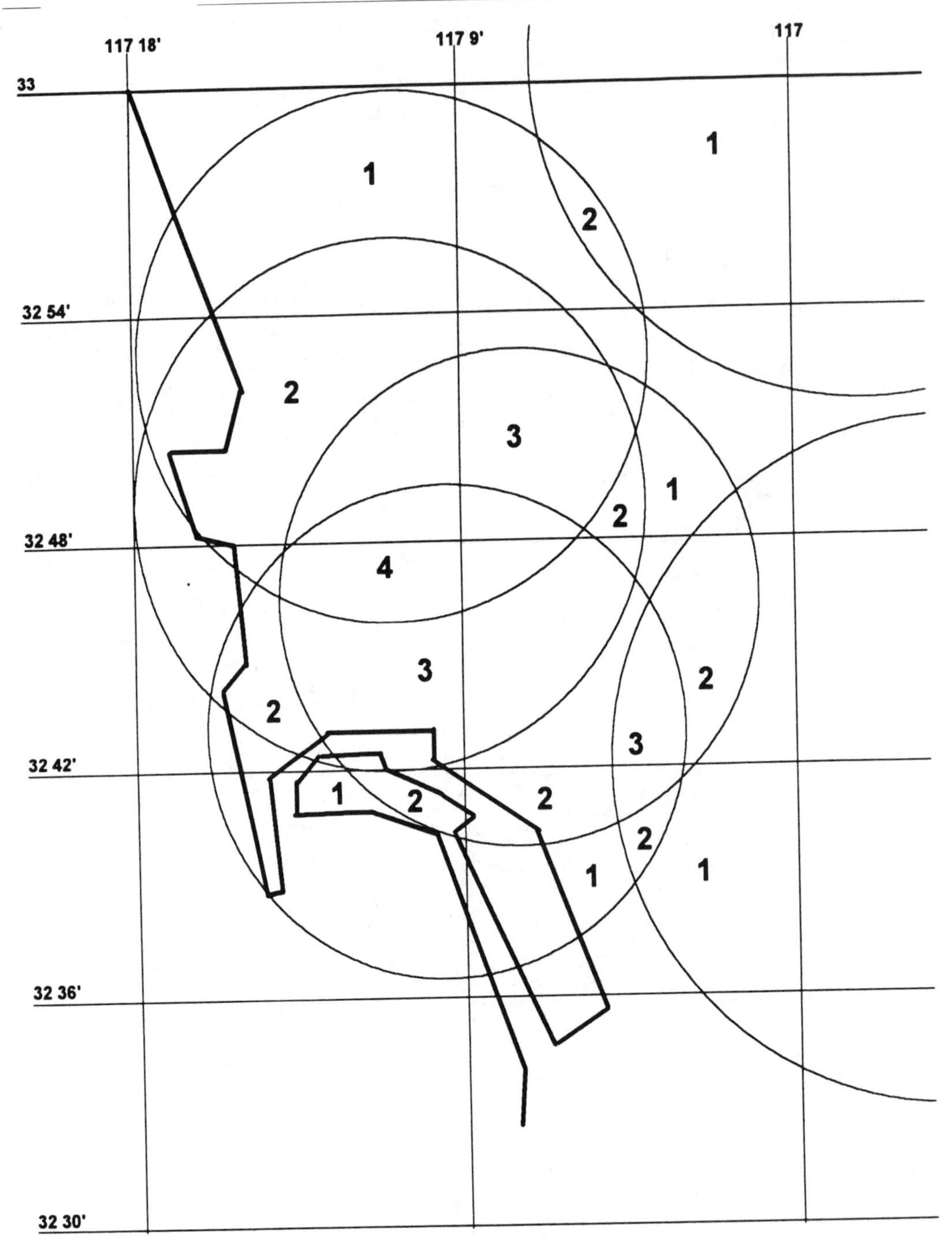

Figure 10-5 *ARDIS Gap Filler Coverage: San Diego*

Cellular coverage is not as easy to depict. PacTel Cellular has 62 base stations licensed in San Diego County;[11] US West has 60. In the geographic area with base station centers between 33° and 32°34′ North latitude and 117°17′ and 117° West longitude, each cellular carrier has ~32 cells. This compares with 6 ARDIS stations. Crudely, cellular deploys 6 to 8 times as many base stations as ARDIS, roughly comparable to nationwide totals. RAM comparisons were not possible; at this writing—August 1993—RAM was not yet licensed in San Diego.

In the core area where ARDIS has quadruple coverage, PacTel Cellular (an arbitrary choice) has seven base stations (see Table 10-2).

Table 10-2 *PacTel Cellular Coverage in ARDIS Intensity Zone*

Address	Latitude	Longitude
Bayard and Garnet	32° 47′ 48″	117° 15′ 8″
2625 Ariane Drive	32° 49′ 35″	117° 13′ 43″
Mt. Ada Rd. & Mt. Rias Place	32° 49′ 11″	117° 10′ 19″
4777 Mercury Street	32° 49′ 41″	117° 8′ 51″
4010 Hicock Street	32° 45′ 27″	117° 12′ 30″
591 Camino de la Reina	32° 45′ 55″	117° 9′ 39″
9656 Yolanda Street	32° 47′ 26″	117° 6′ 59″

In Figure 10-6 the cellular coverage circles were manually plotted on an expanded view of the core (ARDIS coverage in dotted lines) to fit without excessive overlap. There is virtually no doubt that cellular will cover every street-level inch of this core area.

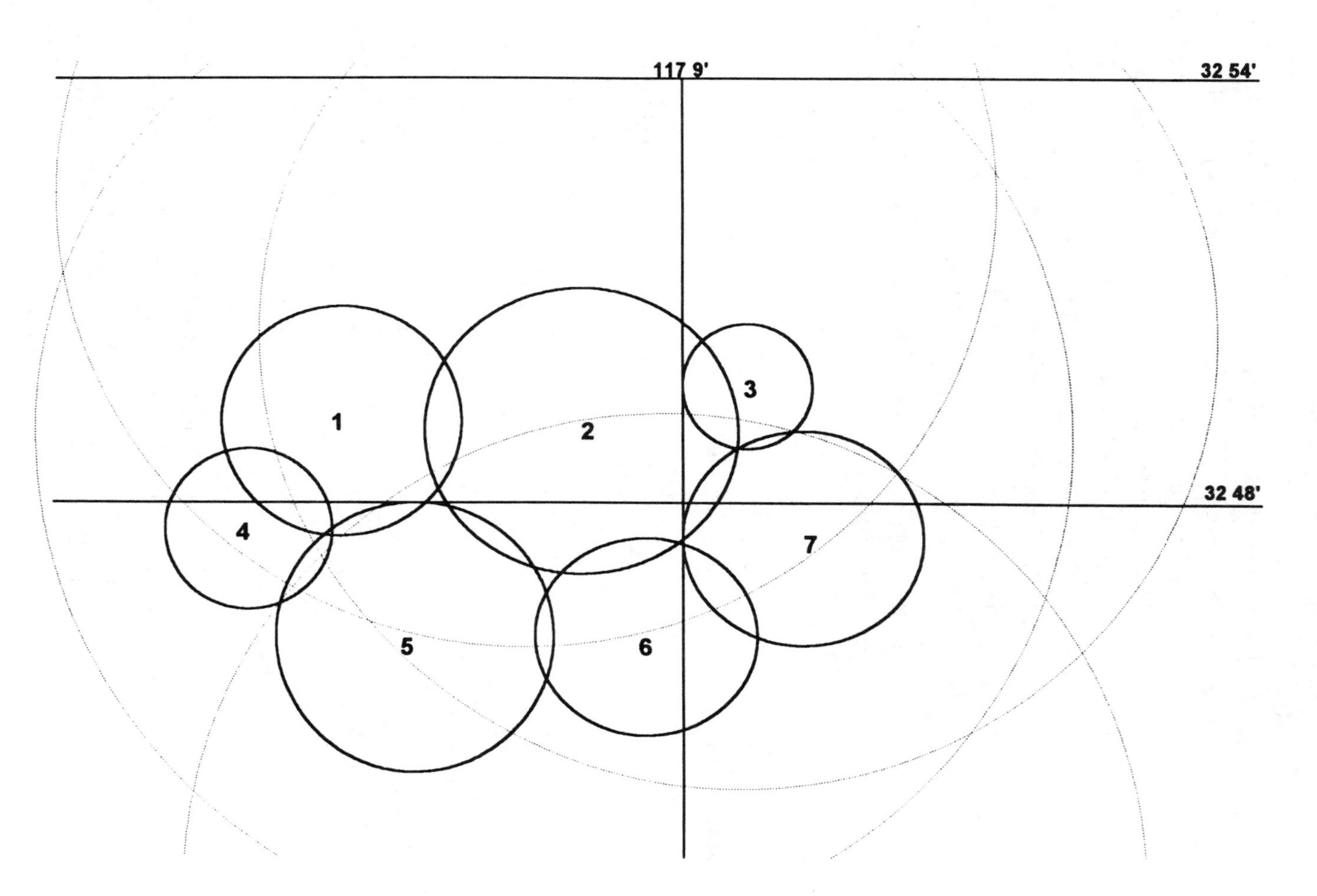

Figure 10-6 *ARDIS versus PacTel Cellular: Intense Coverage Zone*

However, as a handheld user retreats deeper into the depths of a building the restriction in base-station alternatives is likely to extract a coverage price. A simple example: Assume a user is walking down the exterior hall of a building with a clear line of sight through glass to the base station, easily meeting the cellular goal of 90 percent coverage in flat terrain.[12] During the transmission sequences the user turns the corner and moves down a concrete corridor, entering the 75 percent coverage area, not at all unlikely as this is a normal "hilly terrain" target[13] for cellular (actual conditions might be far worse). Message retransmission is required, and base-station access may be a temporary problem.

In ARDIS this transmission might be "heard" on three alternative base stations. Assume the best station is totally blocked, and the three alternatives each have only a 50/50 chance of picking up the transmission. Then $(1 - (.5)^3) = 87.5$ percent of the time the message will be usefully heard somewhere. Orders of magnitude difference? No. Just degrees, improved percentages, with the balance likely in favor of ARDIS *in this critical zone.*

Thus, the application choice: If coverage must be street-level ubiquitous, and occasional dead spots in buildings are acceptable, cellular is the best choice. If the business activity is concentrated geographically and takes place mostly in the interior of buildings, ARDIS is probably a better solution.

10.3.3 ARDIS Capacity Additions

On a nationwide basis ARDIS has ample capacity in place. But capacity shortages rarely occur on a nationwide basis. This is a city-by-city phenomonon. Bartlesville, Oklahoma, is probably set for the remainder of the decade. New York City is a planning problem predating ARDIS: DCS had to continually monitor its load.

The ARDIS approach to additional capacity is fivefold:

1. Acquire frequency pairs, one pair at a time, in cities where load problems are forecast. With this step-by-step process six channel pairs have been secured in Los Angeles and New York City; five in Chicago, Houston, Philadelphia, San Francisco, and Washington; four in Boston, and so on.[14]
2. Deploy separate, independent *base station* hardware for each of these unique channels. These base stations may be configured for special purposes: two channels in New York City are dedicated to street-level coverage, for example.
3. Use synthesized modems to search across the channel pool. The modem problem is more difficult than the cellular case. Channels are not predictably blocked together, and assignments can vary from city to city.
4. Improve the protocol efficiency by combinations of higher bit rates, improved efficiency, better inbound contention techniques, and so on. The RD-LAP protocol can deliver 5 to 6 times the traffic of its MDC4800 predecessor, for example.
5. Use multiprotocol modems to permit the subscriber unit to operate with both old and new protocols.

The resultant infrastructure resembles a wedding cake as shown in Figure 10-7.

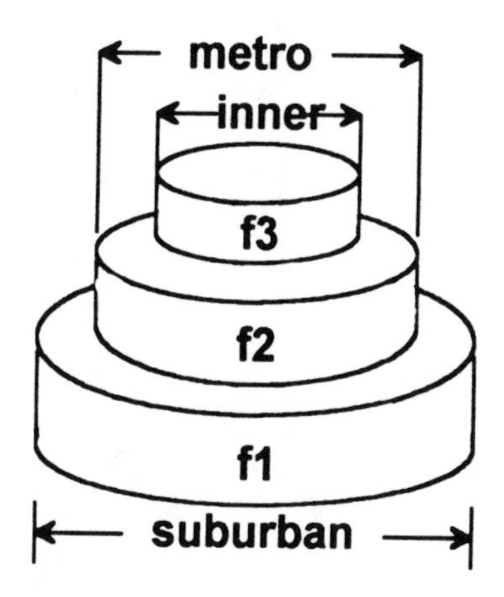

Figure 10-7 *ARDIS "Wedding Cake" Capacity Additions*

REFERENCES

[1] Benjamin L. Scott, EVP and CEO, Bell Atlantic Mobile, *Telephone Week*, 6-21-93

[2] *Mobitex System Description*, 1551-A 296 5073, Section 9.2.1, p. 33.

[3] Ibid, Section 9.2.2, p. 34.

[4] Ibid, Section 9.2.3, p. 34.

[5] John Krachenfels, RAM Director of Business Development, as reported in *Industrial Communications*, 9-28-90.

[6] *Ericsson C719/M Radio Modem Technical Specifications*, 1990.

[7] Tom Berger, ARDIS Vice-President of Radio Network & Product Technology, Lexington Conference, 7-27-93.

[8] Jack Blumenstein, ARDIS President, as quoted in *Mobile Data Report*, 1-27-92.

[9] ARDIS Lexington Conference presentation, 7-27-93.

[10] From FCC radio station licenses.

[11] FCC licenses, 12-17-92.

[12] William C. Y. Lee, *Mobile Cellular Telecommunication Systems*, Section 1.5.2, p. 10.

[13] Ibid.

[14] April 1, 1993, status as reported by ARDIS.

CHAPTER

11

AIRLINK CAPACITY ESTIMATES

11.1 TECHNICAL APPROACH

The number of variables in a packet switched data radio system are virtually endless. This does not mean the task of estimating user capacity is hopeless. Directionally accurate work is possible, and is ongoing in the case of CDPD with its frequent system specification updates. The results are separately available from JFD Associates[1] and are listed here for reference:

1.	MDC4800 Capacity	August 1991
2.	RD-LAP Performance Analysis, Pass 3	June 1992
3.	CDPD V1.0 Airlink Capacity Analysis	December 1993

There are, naturally, unique tasks for each protocol, but the general attack on the problem uses the following sequence:

1. The underlying protocol efficiency, both outbound and inbound, is calculated for a range of message lengths. These figures reflect "normal," ongoing data transmis-

sions: Addresses are compressed, control traffic is ignored, and so on. In the case of discontinuous transmission (as for inbound messages), the unavoidable transmitter turn-on delays, as well as the busy hang time, are also counted. The total bits required to be transmitted for each message length are converted to time intervals.

2. The maximum message-per-second rate is calculated for the *error-free* case.

   Each message is assumed to have a confirming ACK in the opposite channel pair. Further, utilization limits are placed on queued (outbound) traffic, and the appropriate collision formulas, with the necessary limiting assumptions, are placed on contending inbound traffic.

3. The message-per-second rate under error conditions is calculated at a specific target speed and carrier-to-noise (C/N) value.

   For similar protocols/speeds (i.e, RD-LAP and CDPD) the target speed is 30 mph with an average C/N of 20 dB. Key recovery techniques, such as the type of ARQ employed, are exploited.

4. The maximum potential is reduced still further by overhead associated with the particular implementation.

   In the case of CDPD this is channel hopping, for which voice traffic models were written. RD-LAP would reflect the switch from dedicated to keyed output transmissions.

5. The message-per-second rates are then beaten against a variety of average message arrival rates to establish the number of supportable active users.

   Although any interarrival rate may be calculated, the four standard tables include a message rate of one per 15, 30, 45, and 60 minutes. The subscriber-to-active-user ratio is not calculated.

11.2 REPRESENTATIVE SINGLE-BASE-STATION RESULTS

The full reports are needed to understand the intricacies of each protocol. However a simple, relatively argument-free comparison can be demonstrated here.

Assume a case of dedicated, continuously keyed outbound channels. That is, CDPD doesn't hop and MDC4800 and RD-LAP are not used in the ARDIS manner. Message lengths from 5 to 240 octets (the MDC4800 limit), in 5-octet steps, are calculated. The outbound channel utilization is held to 70 percent.

The theoretical error-free message-per-second rate for all three protocols is shown in Table 11-1.

Clearly MDC4800 is now outclassed by the more modern protocols.

Further, the MDC4800 short-message capability is obvious. With a message only 5 octets long CDPD, with four times the raw transmission speed, has only a 3.5 to 1 advantage. This is because MDC4800 is very granular; CDPD is carrying many unused bits at this short length. Move that user message to 240 octets and the advantage jumps to 5.4 to 1.

Table 11-1 *Message per Second: Dedicated Outbound Channel with Retries (70% utilization)*

User octets	Msg/sec MDC4800	Msg/sec CDPD	Thruput improved	Msg/sec RD-LAP	Thruput improved
5	4.56	16.00	3.6	22.32	5.0
10	3.95	16.00	4.2	18.16	4.7
15	3.49	16.00	4.7	18.16	5.4
20	3.13	16.00	5.3	18.16	6.0
25	2.83	10.67	3.9	15.31	5.6
30	2.83	10.67	3.9	15.31	5.7
35	2.59	10.67	4.3	13.23	5.4
40	2.38	10.67	4.7	13.23	5.9
45	2.21	10.67	5.1	11.65	5.6
50	2.06	10.67	5.6	11.65	6.1
55	1.92	10.67	6.0	11.65	6.5
60	1.92	8.00	4.5	10.40	5.9
65	1.81	8.00	4.8	10.40	6.3
70	1.71	8.00	5.1	9.40	6.0
75	1.61	8.00	5.5	9.40	6.4
80	1.53	8.00	5.8	9.40	6.8
85	1.46	8.00	6.1	8.57	6.6
90	1.46	8.00	6.2	8.57	6.6
95	1.39	6.40	5.2	7.88	6.4
100	1.33	6.40	5.5	7.88	6.7
105	1.27	6.40	5.7	7.29	6.5
110	1.22	6.40	6.0	7.29	6.9
115	1.17	6.40	6.3	7.29	7.2
120	1.17	5.33	5.3	6.78	6.7

Table 11-1 *Message per Second: Dedicated Outbound Channel with Retries (70% utilization) (continued)*

User octets	Msg/sec MDC4800	Msg/sec CDPD	Thruput improved	Msg/sec RD-LAP	Thruput improved
125	1.13	5.33	5.5	6.78	7.0
130	1.09	5.33	5.8	6.34	6.9
135	1.05	5.33	6.0	6.34	7.1
140	1.01	5.33	6.3	6.34	7.4
145	0.98	5.33	6.5	5.95	7.3
150	0.98	4.57	5.6	5.95	7.3
155	0.95	4.57	5.8	5.61	7.2
160	0.92	4.57	6.1	5.61	7.4
165	0.89	4.57	6.3	5.30	7.3
170	0.87	4.57	6.5	5.30	7.6
175	0.84	4.57	6.7	5.30	7.8
180	0.84	4.57	6.8	5.03	7.5
185	0.82	4.00	6.1	5.03	7.7
190	0.80	4.00	6.3	4.78	7.6
195	0.78	4.00	6.6	4.78	7.8
200	0.76	4.00	6.8	4.78	8.1
205	0.74	4.00	7.0	4.56	8.0
210	0.74	4.00	7.0	4.56	8.0
215	0.72	4.00	7.3	4.36	7.9
220	0.70	3.56	6.7	4.36	8.2
225	0.69	3.56	6.9	4.17	8.0
230	0.67	3.56	7.1	4.17	8.3
235	0.66	3.56	7.3	4.17	8.5
240	0.66	3.56	7.3	4.00	8.2

RD-LAP appears even better than CDPD and, strictly speaking, that is true. But RD-LAP is steadily losing ground as the message grows longer. At 5 octets CDPD has only 3.5/4.9 = 71 percent of the capability of RD-LAP (once again granularity plays a key role); at 240 octets the ratio has improved to 5.4/6.1 = 89 percent. This can be seen in Figure 11-1.

As the message grows ever longer the advantage switches to CDPD.

Not shown here is the inbound message-per-second rate. CDPD is an excellent inbound protocol but different G levels were employed in the protocol comparisons. Their use would only lead to unproductive confusion here.

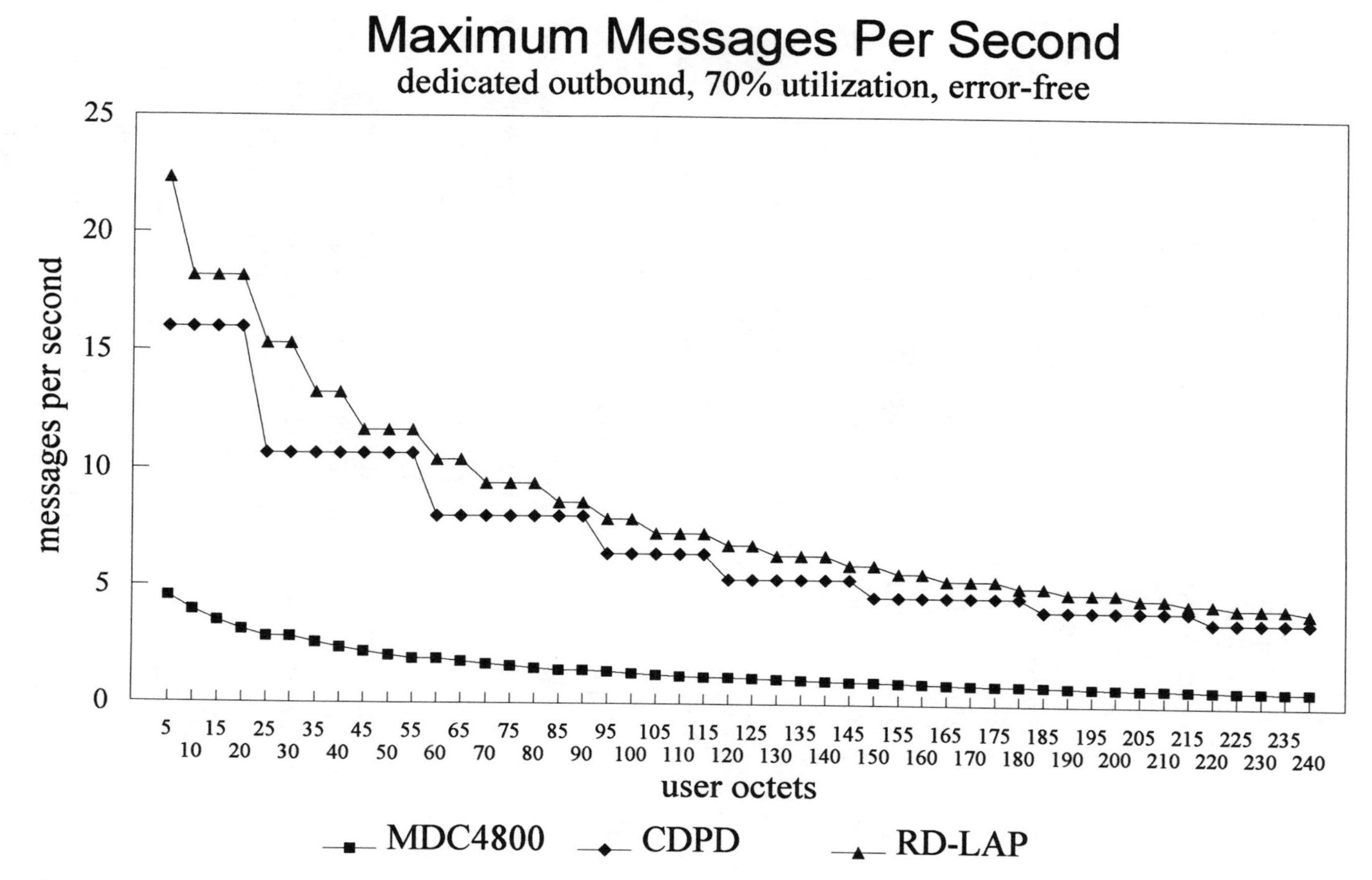

Figure 11-1 *Maximum Messages per Second: Motorola versus CDPD*

11.3 MULTICELL CAPACITY

The capacity calculations for a single base station are of little value in large metropolitan areas with scores of "cells." For example, ARDIS has 34 base stations covering greater Chicago and Ameritech is on a path to bring up 200 CDPD stations in roughly the same geographic area.[2] Assume one were to pick a message length in which RD-LAP and CDPD had identical performance characteristics on a single base station. It would be a grave mistake to say that the CDPD capacity is 200/34 = 5.9 times ARDIS at that equivalent length.

Both systems will be plagued by the "busy cell factor"[3]: the ratio of the percent of traffic carried by the busy cell to the percent of traffic carried by a "normalized" cell. Normalized cell traffic assumes that every cell in the system carries equal traffic. Clearly this is not so.

ARDIS will lose further capacity because of its keyed transmission, single-frequency reuse deployment. CDPD will lose capacity due to channel hopping.

How does one compare the two? On a city-by-city basis with common assumptions concerning message lengths, target speeds, C/N environment, and so on.

An example using the now familiar San Diego area may prove instructive. ARDIS deploys 8 base stations in all of San Diego County. Each of the cellular carriers has ~61. In a more confined geographic area ARDIS has 6 stations versus 32 for each of the cellular carriers. This 6 versus 32 will be the basis for comparison.

Four of the six ARDIS base stations operate nearly as one. It is quite possible (even normal) for, say, base 1 and base 4 to be transmitting simultaneously; we will be punitive and say that each of these stations is unavailable 75 percent of the time because of the need to protect subscriber units in the core area from mixed transmissions. Stations 5 and 6 are virtually independent; we will be punitive again and say that they do interact with the core area 25 percent of the time. Thus ARDIS has 6 base stations that are deliberately silenced much of the time. The effective base-station capacity remaining is 2.5.

The effective base-station capacity for both ARDIS and CDPD will be reduced further by the busy cell factor. In low (~8) "cell"-count configurations average busy-cell factors of 2.7 have been reported by the carriers;[4] in moderate (~28) cell-count configurations average busy-cell factors of 3.8 are normal. These values reasonably match diverse base-station counts.

Thus, the equivalent base-station values are adjusted downward and at different rates. The CDPD deployment will yield about 3.8 times the equivalent base-station capacity as ARDIS in spite of the 5.3 to 1 difference in physical counts. This difference is summarized in Table 11-2.

Table 11-2 *Peak Equivalent Base Stations: ARDIS versus CDPD*

Infrastructure Assumptions	ARDIS	CDPD
Base Stations:		
Installed	6.0	32.0
Equivalent	2.5	32.0
Unbalanced Load:		
Busy-Cell Factor	2.7	3.8
Peak Traffic Busiest Cell*	45%	11.875%
Peak Equivalent Base Stations	2.2	8.4

*CDPD example: $\frac{\text{\% traffic in busy cell during peak hour}}{100 \div \text{number of cells in system}} = 3.8$

$\frac{\text{\% traffic}}{100 \div 32} = 3.8 \qquad \frac{11.875}{3.125} = 3.8$

How much traffic can each of the equivalent base stations carry? The airlink capacity analyses can help with sensible judgments. An alternative way of thinking about message length is to express it in blocks instead of octets. Curves can be displayed for the average number of blocks per second that can be transmitted with a given message length of block size x (see Figure 11-2).

RD-LAP is granular (12 octets per block). With outbound message lengths of 140–190 octets (12–16 blocks per message) the maximum number of blocks per second can be transmitted: ~58. With shorter messages the protocol is still amortizing overhead; with longer messages the retry rates begin to extract their toll. Inbound message lengths of 200 ± 20 octets (17 ± 2 blocks) provide the best inbound yield: ~30 blocks per second. A regression line suggests that 52.4 blocks per second outbound and 27.6 blocks per second inbound are reasonable figures for an average ARDIS cell.

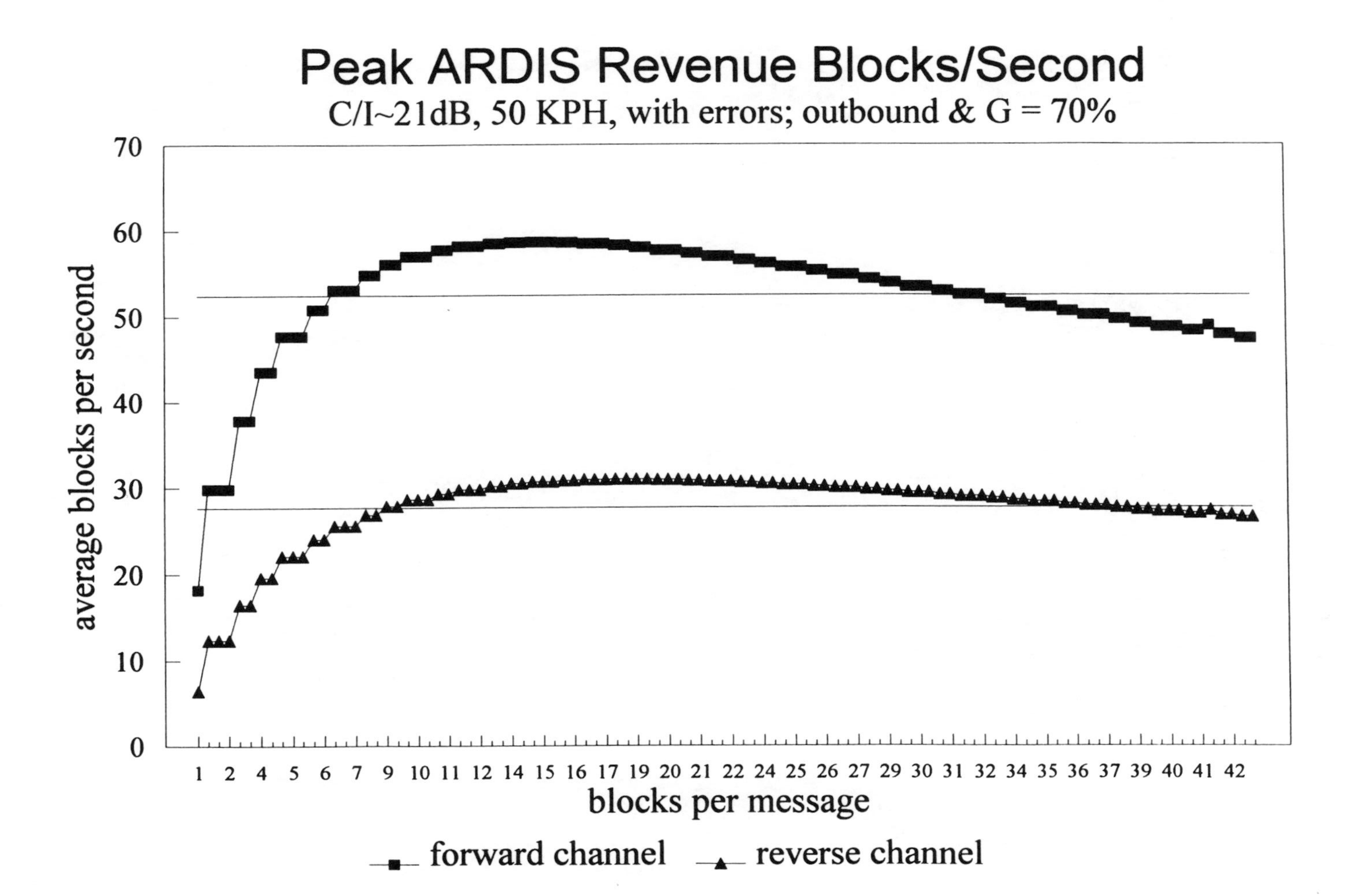

Figure 11-2 *Peak ARDIS Revenue Blocks per Second*

In like manner a working figure for CDPD can be calculated as shown in Figure 11-3.

CDPD is less granular than RD-LAP and has a wave apearance because of the positive impact of selective ARQ on 120-octet boundaries. The regression line yields 19.3 larger outbound blocks per second for CDPD and 10.2 blocks per second for the inbound side.

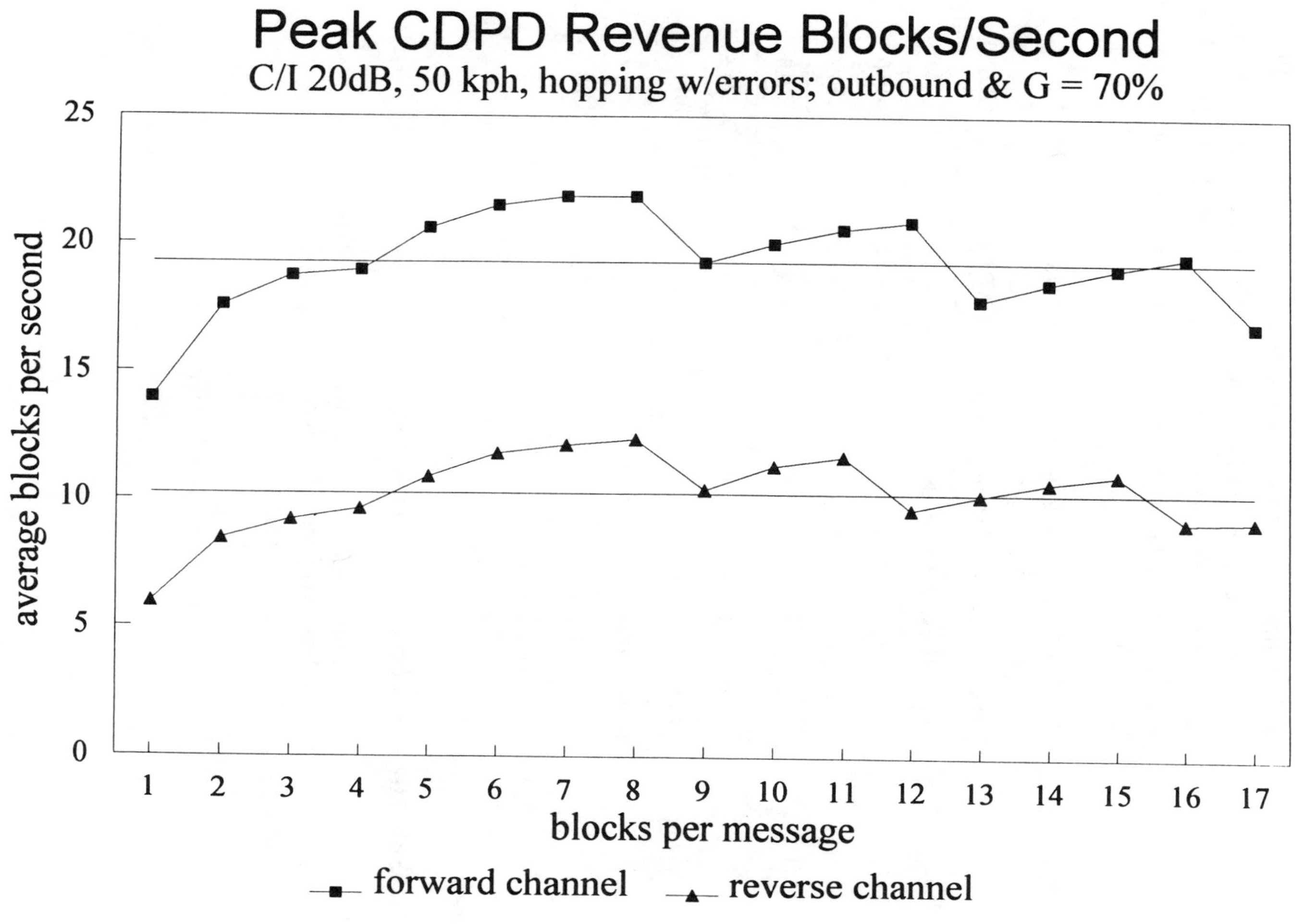

Figure 11-3 *Peak CDPD Revenue Blocks per Second*

This estimating process suggests that, on average, CDPD will deliver about 5 percent more traffic per equivalent base station than ARDIS. The results are summarized in Table 11-3. Thus the San Diego capacity comparison yields the data shown in Table 11-4.

Table 11-3 *Traffic Capacity Assumptions: ARDIS versus CDPD*

Traffic Capacity Assumptions	ARDIS	CDPD
Peak Blocks per Second:		
Forward channel	52.4	19.3
Reverse channel	27.6	10.2
	80.0	29.5
Optimum block granularity	12.0	34.25
Octets per second	960	1010
User TRIB (Effective bps):		
Forward channel	5030	5288
Reverse channel	2650	2795

Table 11-4 *San Diego Capacity Comparison: ARDIS versus CDPD*

	Equivalent Base Stations		Octets per Second		Traffic Potential
ARDIS	2.2	×	960	=	2112
CDPD	8.4	×	1010	=	8484

(Note: this is ~25 "faxes"/minute)

CDPD will have four times the capacity of ARDIS in San Diego if 5.3 times as many base stations are deployed. The key question: Is the investment worth it?

REFERENCES

[1] JFD Associates, 46 Saddle Rock Road, Stamford, CT 06902; voice: (203) 356-1889; fax: (203) 978-1876; CompuServe (also via Internet & RadioMail): 73260,2036.

[2] JFD Associates meeting at Ameritech, 7-14-93.

[3] Compucon Analysis of Spectrum requirements of the Cellular Industry, p. 12.

[4] Ibid, p. 13.

CHAPTER

12

SYSTEMS AND SUBSYSTEMS

12.1 APPROACHES

The airlink subsystem is defined as the span between the user host's connection point and the subscriber unit. It includes the service provider's message switch/area controller, base stations, and the radio modem. There are (at least!) two philosophical *control* approaches:

1. Decentralized, where the subscriber unit, assisted by a high-function base station, makes many channel assignment decisions. A primary example of this approach is the Ericsson system (Mobitex) used for RAM. This also describes CDPD in a dedicated channel configuration.
2. Centralized, where the communication control portion of the subscriber units, as well as the base station, are very low function. Here higher level processors control all ele-

ments of channel assignment and message sequencing for a given area (which can be multi-state). A primary example of this approach is the Motorola system(s) used for ARDIS.

12.2 RAM

For maximum differentiation with ARDIS the following scenario is used. Subscriber units are:

1. Vehicle based (this is *not* a hard requirement; the Mobidem is clearly a portable device)
2. Roaming at least area- or citywide
3. Served by a pool of frequencies arranged in a cellular-like configuration
4. Gathering and measuring signal strengths from multiple surrounding base stations to select the "best" base station for their purposes

A simplified view of the Mobitex network hierarchy is shown in Figure 12-1.

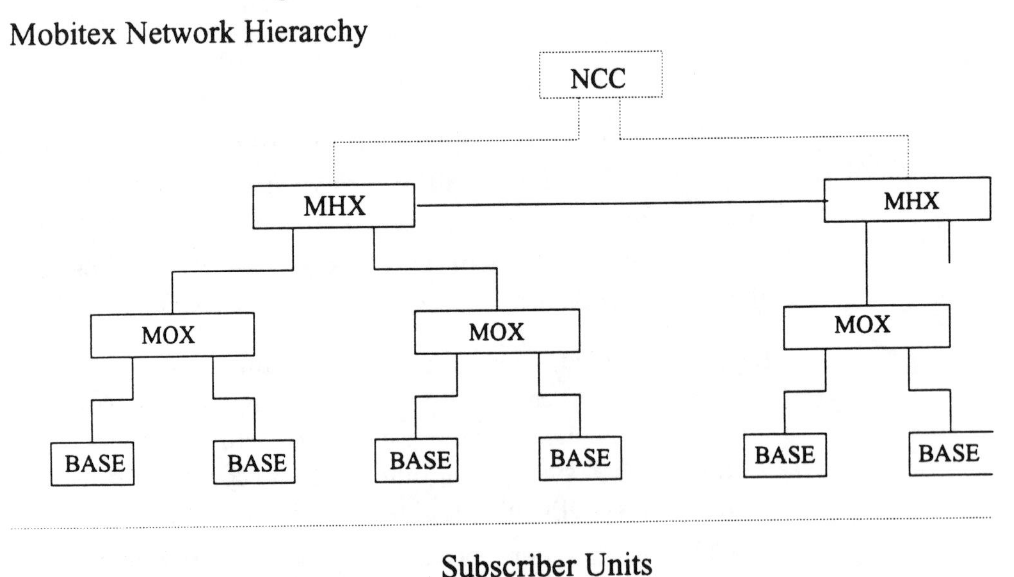

Figure 12-1 *Mobitex Network Hierarchy*

The subscriber units communicate with high-function base stations; the base stations are capable of controlling the connection to other subscriber units registered to the same base without higher-level reference.

If the target is a fixed unit in the same area, or a subscriber unit not registered to this base, the message flows to the area exchange: the MOX.

If the traffic is out-of-area, the message flows still higher to a main exchange: the MHX.

The Network Control Center (NCC) does not take part in traffic handling; it includes operation and maintenance functions as well as the subscription handler.

Control of this hierarchical system depends upon the allocation of national system channnels; their existence insures that the mobile need not scan far for control instructions. Message traffic normally flows over one of four types of traffic channels.

The degree to which functions are decentralized is demonstrated with the following example:

1. The subscriber unit is switched on and its synthesizer begins to scan the frequency pool.
2. Roaming signals from surrounding base stations are gathered by the subscriber unit and their field strengths are measured.
3. The subscriber unit chooses a base station based on this evaluation.
4. If the selected base station is the same one last used, the network is ready; if not, a roam packet is generated by the subscriber unit and:
   a. The roam packet is transmitted on the National System Channel.
   b. If the base has no subscriber data, a connection request is transferred to the superior node, the MOX.

c. Assuming the MOX is able to obtain subscriber information, the new base is updated with the necessary information (this process can be extended to the MHX as well)

d. The network is now ready; everybody knows the whereabouts of the SU.

Personal experience with RadioMail reveals that this search interval can be many seconds, comparable to cellular dial time.

Availability numbers for RAM are difficult to find (unlike ARDIS, which trumpets their results monthly). A "network availability" number of 99.94 percent has been seen, but without clear explanation as to what the number covers. For example, the cellular-like configuration can cause a "hole" in local coverage if a single cell is out of service. This hole will persist for the hours necessary for the repair crew to be dispatched and complete the service call. Thus, it is unlikely that individual cell failures, though locally disruptive, are counted in the network availability statistics.

Assuming the network availability number is accurate and describes some higher level of failure, the RAM network is out of service 26 minutes in each 30-day month.

12.3 ARDIS

12.3.1 System Approach

For maximum differentiation with RAM, subscriber units are assumed to be:

1. Pedestrian handheld
2. Moving over at most a city—and often just between buildings/floors of buildings, deep within the walls

3. Served by only one frequency pair, arranged in an adjacent-cell, diversity-reuse configuration
4. Receiving and transmitting messages without any physical knowledge of base-station location or signal strength

A simplified view of the ARDIS network hierarchy is shown in Figure 12-2.

ARDIS Network Hierarchy

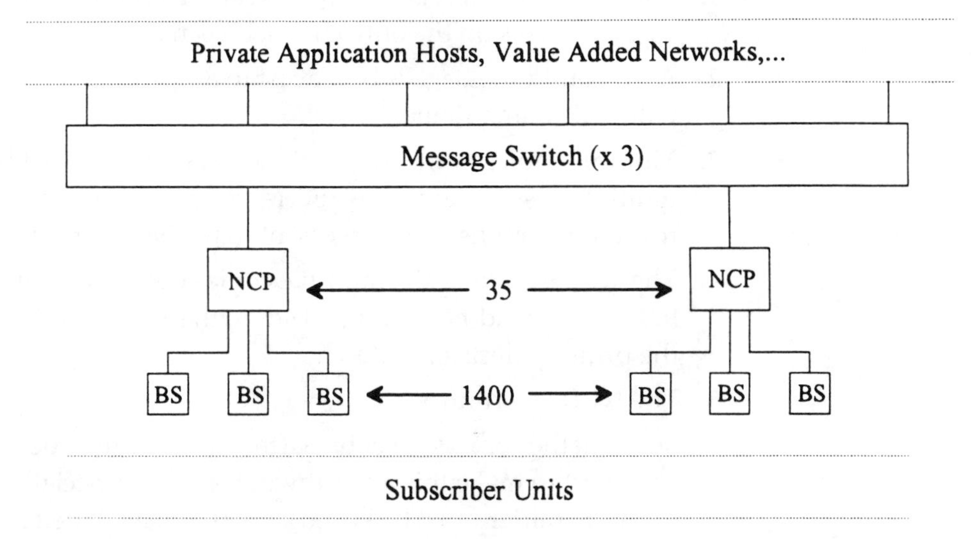

Figure 12-2 *ARDIS Network Hierarchy*

Subscriber units communicate with low-function base stations that act as relay points to the network communications processor (NCP). Here all decisions are made as to which transmitter to key (and when), to set busy or not, and so on. While NCPs are regionalized for control and backup, many service a single city. Until recently the subscriber unit working in Chicago was not really expected to show up in New York City. ARDIS's "blue-collar" subscriber base were not big roamers. Executive-class users, particularly those using e-mail, are forcing useful changes here.

There is no concept of a control channel. All control rests in the NCP. The single allocated channel is used only for data transmission.

The degree to which functions are centralized can be demonstrated with the following pre-frequency agile (still the bulk of the subscriber base) example:

1. The subscriber unit is switched on and its crystal-cut oscillator fixes on the only channel it can hear.
2. An inbound message is broadcast to all base stations within listening range.
3. Multiple copies of the inbound message are received by multiple base stations; the FEC is stripped away; the resulting wireline message is sent to the NCP for action.
4. The NCP evaluates the multiple copies discarding all but best and second-best. A signal strength indicator (SSI) is the primary decision criteria.
5. The NCP determines:
    a. Whether or not to set busy (the message may be too short to make this a worthwhile action, especially if a transmitter must be keyed); if yes, the NCP determines on which base stations busy should be set.
    b. Which is the best path to reach the subscriber unit:
        1. If a long outbound queue exists on the primary path, the alternate can be tried.
        2. If the SSI indicates that the subscriber unit is in a fringe area, the NCP may queue the message until it can complete all transmissions on adjacent channels that might be detrimental to packet delivery.
        3. If the SSI is robust, multiple outbound messages—all using the same frequency—may be keyed on adjacent channels.

12.3.2 System Details

The 1400 ARDIS base stations provide coverage in 400 geographic areas. Perhaps 300 of those areas are one station diameter:

Bridgeton, NJ	Lakeland, FL
Bartlesville, OK	Lincoln, NE
Coalinga, CA	Manchester, NH
Concord, NH	Ponca City, OK
Fargo, ND	Provo, UT
Grand Forks, ND	Sioux City, IA
Kokomo, IN	etc., etc., etc.

The 1,100 remaining are focused on the top 100 metropolitan areas with wide ranges in the number of base stations per area (already mentioned: Chicago at 34, San Diego at 8).

Ignoring redundancy, ~29 operational network control processors (NCPs) service these 1,400 base stations. The wireline connection between base station and NCP is T1/T3 digital service arranged as shown in Figure 12-3.

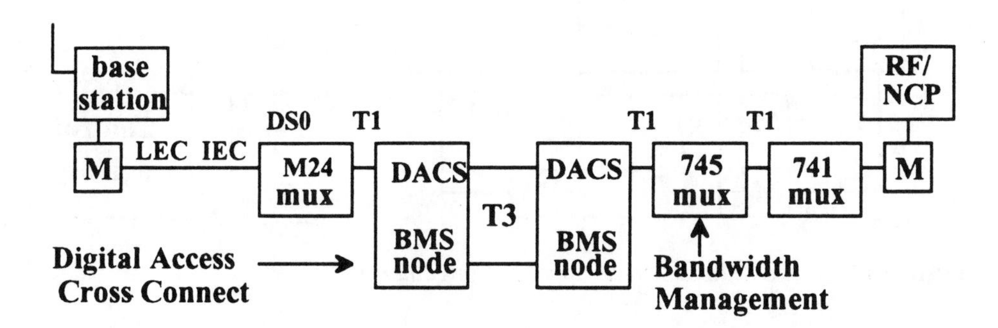

Figure 12-3 *ARDIS Base Station to RF/NCP Wireline Configuration*

With redundant units the 35 NCPs, formally scattered across the nation, are now grouped in six licensed space arrangement (LSA) AT&T hardened facilities, with high-security limited access, uninterruptible power, and backup generators:

1.	Atlanta	5
2.	Chicago	6
3.	Dallas	5
4.	Los Angeles	6
5.	Washington, DC	7
6.	White Plains, NY	6
		35

NCP redundancy at each LSA is achieved as shown in Figure 12-4.

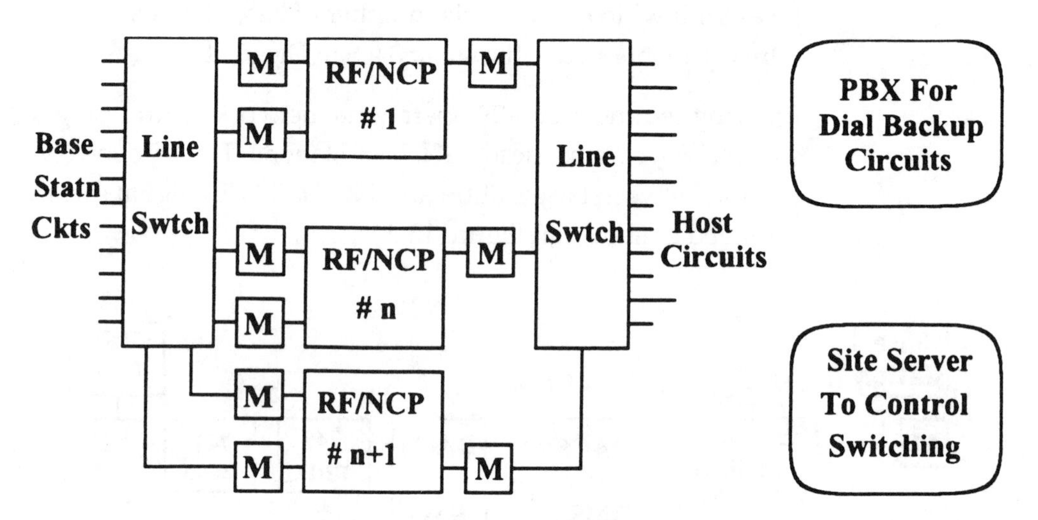

Figure 12-4 *ARDIS Licensed Space Arrangements (LSAs)*

Unlike cellular configurations, a single base station out of service in a large metro area is often unnoticed by the user. If, say, triple coverage exists the user may have to turn toward a window with the handheld upon base-station failure, but the message gets through.

The LSAs are interconnected to three network switching centers:

El Segundo, CA
Lincolnshire, IL
Lexington, KY

El Segundo and Lincolnshire provide primary operational control; Lincolnshire carries ~60 percent of the traffic. Lexington is a backup site where development work and billing/accounting are performed. All three centers are connected via dual, diversely routed 56 Kbps Tandem "Expand" links. Dial backup facilities also exist. Disaster situations (California earthquake destroys El Segundo) have been simulated. Total system switchover/recovery is completed in about two hours.

12.3.3 Network Availability

When private wireless packet systems began, moderate attention was paid to uptime. IBM's DCS system, for example, had its area controllers in IBM branch offices where they could be restarted manually—when the branch was open. A failing controller, which took out at least an entire city, could be patched around with a 30 to 40-step dial backup sequence. The basis for high availability was the assumption that the controller would not go down often—which, indeed, it did not. In addition the handheld had a built-in, slow-speed wireline modem for emergencies.

These techniques were not adequate to the demands of a public network. Thus, ARDIS radically upgraded the system to achieve 24-hour/day, 7-day/week coverage. Network availability goals were set at 6 sigma—for all events, scheduled and unscheduled—dramatically higher than the 99.9 percent performance that produced nearly 44 minutes of downtime per month. The new target is no more than 9 *seconds* of downtime in a month.

For the first half of 1992 ARDIS network availability was ~99.985 percent (~6.5 minutes outage) per month. Since that time change teams have been upgrading every base station in the system to accommodate automatic roaming, and overall system availability has declined to 99.97 percent (~13 minutes per month). After the system stabilizes availability is expected to improve. Note that ARDIS contractually guarantees availability to its users.

PART

DEVICES

CHAPTER 13

DEVICE INTRODUCTION

13.1 COMBINATIONS AND PERMUTATIONS

A bewildering array of subscriber unit options are available for the terrestrial data market ranging from fully integrated terminal/modem/radio units to multibox combinations of PCs, faxes, and cellular transceivers. A (too) simple representation of some of the variations is illustrated in Figure 13-1.

This convoluted collection of hardware alternatives is extraordinarily volatile. It is not uncommon for monthly PC magazines to publish detailed reviews of a product only to have key details of the review revealed as obsolete by the vendor's new ad (especially price). Thus an exhaustive review of the wide range of product possibilities makes little sense for a book with a slow publishing schedule.

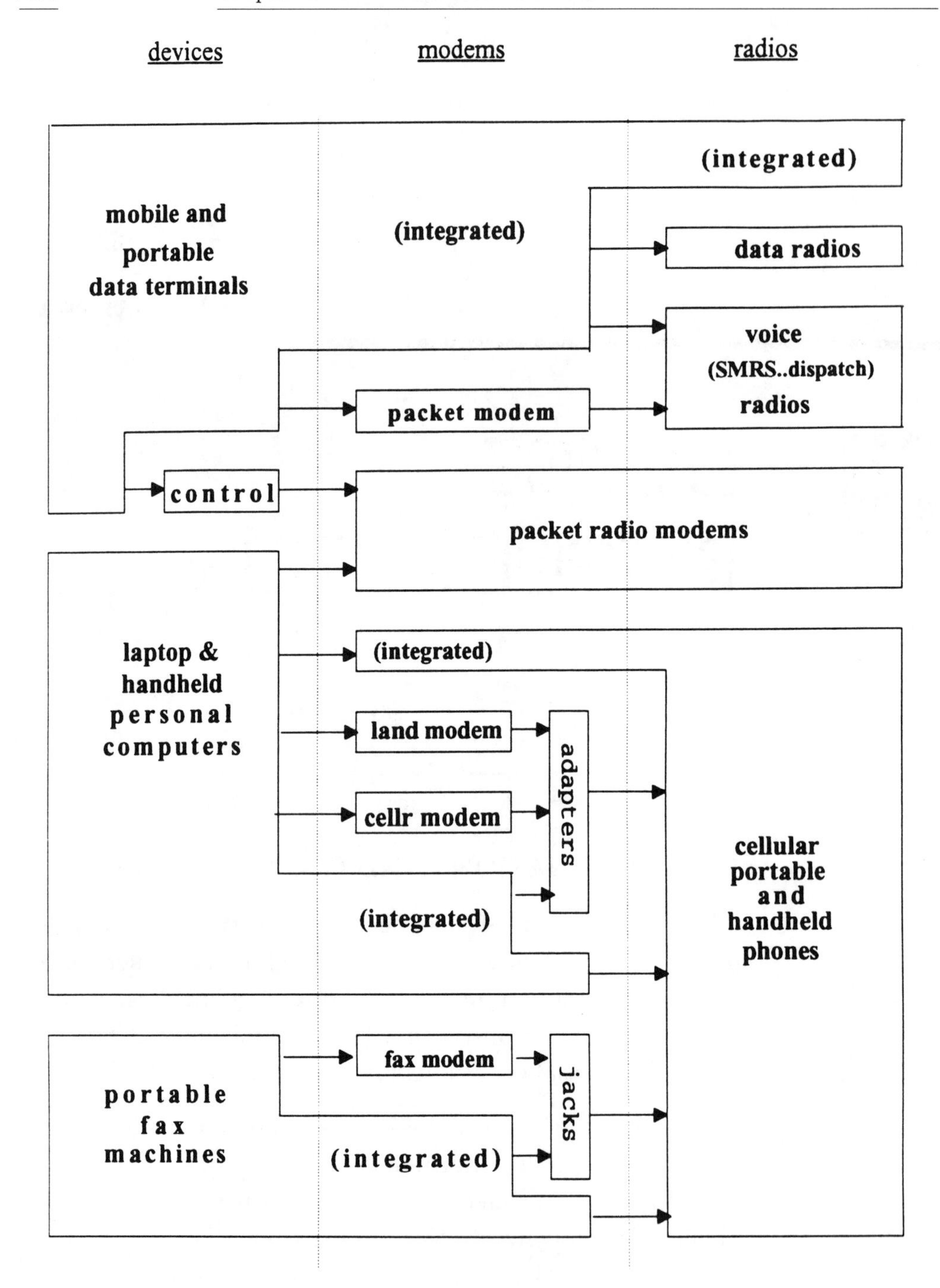

Figure 13-1 *Device Combinations and Permutations*

There is also the question of personal choice in devices. User "A" may balk at carrying a palmtop; user "B" will carry the latest active matrix laptop even if it weighs 8 or 9 pounds with battery chargers and adapters.

However, key trends for pivotal segments that permit the devices to communicate must be examined in order to understand where wireless communication is going. This chapter and the next deal with modems: cellular and packet switched, external and internal, with and without radios or adapters.

First, a look at the soft technologies that are driving modem development: modulation schemes, error detection/correction, and data compression.

The next chapter uses practical examples of existing products to map what appear to this author to be clear communication trends.

13.2 ALPHABET SOUP

13.2.1 Enabling "Soft" Technologies

Thirty seconds into a review of communication product specifications one encounters the war of "vee-dot" (V.) versus MNP. It is unwise to dismiss this sometimes unintelligible gibberish as an unimportant attempt to break/protect vendor proprietary interests. These alphabetic designators are often surrogates for key communication developments required for the success of data radio.

13.2.2 Modulation: V.32/V.32bis

Most packet switched data radio devices have a history of proprietary modulation techniques with arcane names such as "modified duo-binary" (MDI's MMP protocol) to noncoherent baseband FSK (Motorola's MDC4800). In contrast, the modems used on cellular were usually conventional wireline (or wireline derivative) devices with standard CCITT-standardized V series modulation techniques. The most common were:

CCITT Series#	Line Speed	Type	Modulation Technique	Switched Lines	Leased Lines
V.22	1200	FDX	phase shift	yes	pt-to-pt, 2 wire
V.22bis	2400	FDX	QuadAmpl	yes	pt-to-pt, 2 wire

Between 1976 and 1982 IBM's Gottfried Ungerboeck developed the theory, and received the patent,[1] on what is now known as Trellis Coded Modulation (TCM).

Briefly, TCM introduces redundant bits among the information bits to be transmitted. These extra bits are used for detecting and sometimes correcting transmission errors. Of particular importance is the fact that the redundancy is added in such a way that sequential data symbols cannot succeed each other. The possible states can be represented by a transition diagram that (imaginatively) resembles a trellis. The state sequences can be explored recursively using the Viterbi algorithm to predict the most likely sequence. An error that disrupts this sequence can often be corrected by substitution.

The TCM goal was to permit 9600 bps transmission over dial lines. Initially ignored (even within IBM) continued testing, as well as the development of affordable signal processors, proved that TCM was eminently practical. In 1986 the technique was standardized by CCITT as V.32 over the resistance of the U.S. delegates who stated "the V.32 recommendation is not workable . . . and is needlessly sophisticated for most PC communications."[2]

Three years before the formal V.32 standardization, IBM research saw that the TCM techniques could be pushed to higher user speeds.[3] Thus was born V.32bis. In this implementation 7-bit symbols are transmitted at 2400 baud to produce a raw transmission speed of 16,800 bps. The user bit rate is specified as 14,400 bps (14 percent redundancy). The TCM technique was embraced quickly by manufacturers. IBM was beaten to market by the 14,400 bps Codex 2360/2660 in 1984, two years before CCITT endorsed the 9600 bps standard!

With relentless reduction in both size and cost, V.32/V.32bis modems became the favored modulation technique for circuit switched wireline modems in 1992. Because the standard permits fallback to half speed under adverse line conditions (e.g., 9600 bps drops to 4800 bps), this slower speed operation proved useful on cellular as well.

Nearly simultaneously Motorola deployed an enriched TCM to improve the correction power of its RD-LAP packet switched modems. Here 8 bits per baud are sent at a raw transmission speed of 19,200 bps. The user yield is 14,400 bps (25 percent redundancy).

CDPD has chosen not to employ TCM, using GMSK instead. However, the transmission rate remains identical to Motorola's: 19,200 bps is sent to obtain 14,400 bps for user purposes.

13.2.3 Error Detection: V.42/V.42 Fast and MNP4/10

In 1988, after bruising political battles, CCITT recommended V.42 as its error-control standard. The competitive approaches were a protocol called LAP-M and Microcom's MNP4. LAP-M is a variant of the HDLC protocol devised for ISDN: LAP-D. It is a full-duplex protocol and was backed by the U.S. modem manufacturer Hayes, as well as by AT&T and British Telecom.[4] MNP4 is a full-duplex technique developed by U.S. modem manufacturer Microcom and backed by the Belgian, French, and Italian PTTs.[5]

Note that both approaches are speed independent. One does not have to have V.32/V.32bis for V.42. In fact virtually all initial implementations were on the older, slower V.22 modulation standard.

Both techniques are CRC detect and ARQ retransmit. Technical differences are minor; both protocols provide virtually identical performance. The camel compromise: V.42 requires modems to use LAP-M and support a secondary mode of operation that uses MNP4. The theory: Enhancements would flow to LAP-M, gradually retiring MNP4. Microcom did not permit that to happen.

Microcom uses V.42 as a base to extend error detection with a series of MNP levels, some of which are useful for cellular data. The core of the approach is to depend upon a 16-bit CRC for the detection of an error in a packet that does not exceed 256 octets. With MNP4 the size of this packet varies with transmission quality: The better the "line" (reflecting the original design), the larger the packet. Continuous ARQ can be used to retransmit faulty packets.

Collectively, the "MNP cellular" features are called MNP10. When used with cellular (which is detectable by the modem) the key differences with the MNP4 versions are:

1. MNP4 operating on wireline uses large packets at the beginning of a session; if there are too many errors the packet size is reduced until the error rate reaches an acceptable level. MNP10 does the opposite: It begins with small packets and continues to increase packet size in the absence of errors.
2. Bit rates are reduced in the presence of noise; when the noise condition eases, the bit rates are advanced again (this function is not specified in V.42). These speed shifts can be dramatic: 300 bps under adverse conditions to 4000 bps in a "clean" environment.[6]
3. The modem begins transmission at low modulation amplitudes. If channel background noise is detected the receiving modem can ask the sender to increase signal power. In the presence of distortion the receiver can ask the sender to decrease signal power.

Meanwhile, some of the original V.42 LAP-M proponents were far from idle. AT&T Paradyne continued its own proprietary work on V.42 improvements for cellular. Called ETC, "enhanced throughput cellular," the protocol has been tested at Bell Atlantic Mobile, Ameritech, and Southwestern Bell.[7] It is modulation dependent and requires either V.32 or V.32bis as the underlying technique.

As might be expected AT&T Paradyne claims[8] markedly superior performance to MNP10: "ETC blows MNP10 out of the water"; Microcom disputes all the test results. Arguments aside, it is clear that the competition between the two will force superior data performance for cellular.

13.2.3.1 *Maximum MNP10 Throughput*

Assume an MNP10-equipped pair of modems operating in a pristine environment, transmitting a 500-octet user message. MNP cellular starts with small packets (32 octets) and increases packet size until errors "start to significantly affect efficiency." In our fictitious error-free state the message will be sent as seven packets of sizes 32, 32, 64, 64, 128, 128, and 52 octets. While ACKs between segments can be specified, we will assume continuous ARQ (which just fits!) that does not call for a retransmission.

The sequence is shown in Figure 13-2, where the link requests are 8 octets, the ACKs are held to 7 octets, each PDU is surrounded by flags and CRC-16, and the disconnect is 7 octets.

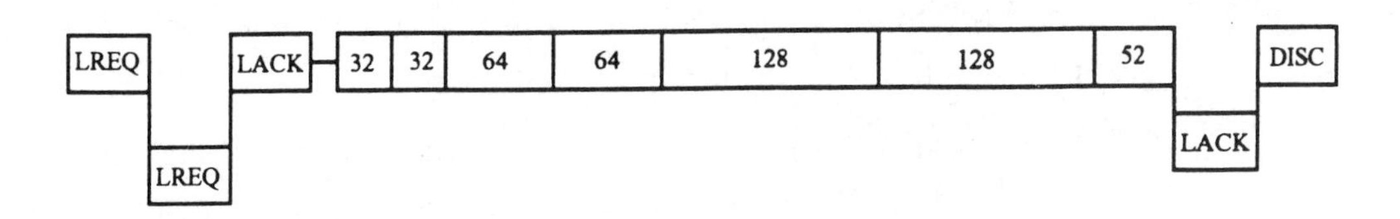

Figure 13-2 *MNP Maximum Throughput*

The total octets transmitted are:

$$8 + 8 + 7 + 35 + 35 + 67 + 67 + 131 + 131 + 56 + 7 + 7 = 559 \text{ octets}$$

This theoretical high efficiency (500/559 = 89 percent) could yield very attractive TRIBs. However, the transmission time at 14,400 bps would be 310 milliseconds. Carrier negotiation can easily be 100 times that long. If high TRIB is a goal, the message must be a long one to overcome the punishing overhead.

Even with long messages, real-world TRIBs half the rated speed of the modem would be pleasing, especially while moving. If the test results noted in Section 14.2 are roughly correct the total elapsed time to send a 100,000-octet file, with 12 percent overhead, would be 57 seconds negotiation plus (800,000 × 1.12) ÷ 14,400 bps = 62 seconds transmission time. These numbers yield a TRIB of ~7530 bps. Note that the ETC equivalent would be 10,540 bps—which was not claimed. Other damaging hits to the message that force packet retransmission are clearly present. Shrinking negotiation times is key to improved performance and lower airtime bills.

13.2.4 Data Compression: V.42bis and MNP5/7

In a reprise of the V.42 battles, CCITT was faced with two competing proposals for modem data compression in 1988.

The first was the extant MNP5 developed by Microcom. This Microcom approach used a real-time adaptive algorithm to achieve a nominal doubling of throughput. That is, under good conditions a 2400-bps modem would be able to make an effective data transfer rate of 4800 bps.

Fully aware of its modest compression achievement, Microcom delivered MNP7 in the second half of 1988. This technique used Huffman coding (few bits for frequent English language letters such as E, T, A; more bits for infrequently appearing letters such as Q, Z) with a predictive algorithm. The result: ~3-to-1 improvement under good conditions.

ACT's CommPressor alternative based on Lempel Ziv techniques that could yield 4-to-1 improvements was selected as the V.42bis standard in September 1989. Microcom continues to license and build its MNP alternatives but admits that V.42bis "will increasingly become the preferred method of data compression."[9]

The actual compression achieved in practice varies widely with the application. The algorithm works by recognizing repeated patterns in data and substituting shorter symbols for them. The more repetition a file has, the greater the compression. But purely random data contains no patterns at all and cannot be compressed. Further, data files that have been encrypted through a randomization process will also show little reduction in size because the data has had identifiable patterns removed. This suggests that CDPD might not benefit from data compression.

A data compression summary[10] is shown in Figure 13-3.

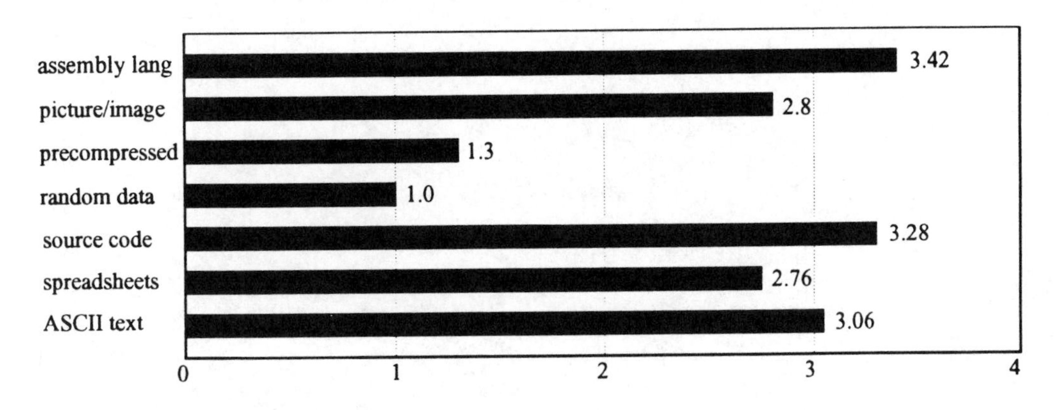

Figure 13-3 *Data Compression Summary*

These are average ratios. A compression ratio of 2.0 indicates that the data can be compressed by a factor of 2 and transmitted in half the time needed to transmit it uncompressed.

REFERENCES

[1] "Fractional Tap-Spacing Equalizer and Consequences for Clock Recovery in Data Modems," *IEEE Transactions on Communications*, August 1976, pp. 856–864; Method and Arrangement for Coding Binary Signals and Modulating a Carrier Signal, U.S. Patent No. 4,077,021, February 1978; "Channel Coding with Multilevel/Phase Signals," *IEEE Transactions on Information Theory*, January 1982, pp. 55–67.

[2] *PC Week*, 1-19-88.

[3] "Proposal for a 14,400 Bits per Second Modem for Use on 4-Wire Telephone Circuits," *IBM Europe: Contribution to CCITT Study Group XVII*, No. 100, February 1983.

[4] *PC Week*, 5-10-88.

[5] *Info World*, 5-30-88.

[6] *The Cellular Handbook*, Microcom publication 10K-RES-6/90.

[7] *Mobile Data Report*, 2-15-93.

[8] AT&T Paradyne as reported in *Mobile Data Report*, 7-5-93.

[9] Microcom, *Microcom Networking Protocol*, 1990 edition, p. 3.

[10] *Byte Magazine*, November 1990, p. 360.

CHAPTER

14

MODEMS AND ADAPTERS

14.1 MODEM TYPES

Modems have four levels of sophistication when operating in the wireless environment:

1. No consideration for errors. This class of modem is usually low speed (300 bps) where error tolerance is high, and control rests with the communications software driving the modem. A (rapidly) dying breed.
2. Use of only error-detection mechanisms, combined with ARQ techniques, and often accompanied by packet resizing to optimize performance. The foremost examples are early Microcom units and their many MNP-licensed cousins. These units are also moving slowly into history.

3. The use of trellis-coded modulation to provide error-correction capability without major impact on efficiency. This more reliable data transmission technique is also combined with ARQ to ensure successful message delivery.
4. The use of powerful error-correction coding techniques, deliberately sacrificing user capacity, in an attempt to deliver a clean message *without* retransmission., This class of modems is seen in early cellular connections made by Spectrum and (later) Millidyne. They occur most commonly in packet switched networks: Motorola's many MDC4800 variations and Ericsson's Mobitex units used on RAM. This approach is also at the heart of CDPD's airlink protocol.

The modems can be packaged as stand-alone units or, increasingly, integrated into terminal devices, laptops, or cellular phones. The Personal Computer Memory Card International Association (PCMCIA) standard, along with years of groundlaying product development, has now catapulted the integrated unit to the forefront.

In either case, if used on cellular, vendor variations on the cellular instruments usually result in an adaptive device or cable between modem and cellular radio. This adapter can be "dumb" or "intelligent." Intelligent adapters mimic wireline dial operations for communications software.

Some modems also contain the radio transceiver, especially packet switched units. Integrated radio modems have been available for a decade at premium prices. PCMCIA developments promise to make this level of integration tractable in 1994. When radio modems are used on cellular the voice handset is usually present for double duty.

14.2 CELLULAR MODEMS

14.2.1 Major Trends

Multiple slow-speed units were announced within a year of the initial cellular rollout in October 1983. Many were failures, but all contributed to the relentless trend toward lower weight and smaller physical volumes that now make integration possible. Each has played a part in advancing bit rates while simultaneously lowering prices.

These trends are difficult to represent graphically because of the collapse of both weight/volume characteristics and prices in the face of enormous increases in raw bit rates—especially with the advent of pocket V.32/V.32bis modems. The approach was to first present stand-alone, non-AC-dependent modems plotted on a log scale so as to discriminate among the more recent product developments. Weight was used as a surrogate for physical trends (see Figure 14-1, next page).

For nearly six years the cellular modem physical trends were pleasant: a 300-bps modem in 1984 weighed 20 ounces; at the end of the period a 1200-bps unit weighed the same 20 ounces and a 300-bps unit weighed 5 ounces. This compound annual weight reduction of ~20 percent seemed a new law of physics.

The 1991 introduction of the Telebit QBlazer broke the trend line. This modem departed from the CCITT V.22 (DPSK)/V.22 bis (QAM) modulation techniques that were clearly topping off at 4000 bps on wireline systems. Telebit employed V.32 Trellis Coded Modulation as well as recovery strategies originally developed for wireline units; with the transmission speed reduced, typically to 4800 bps, these techniques made a good fit for cellular as well. Raw bit rates thus rose by a factor of 8; weights/volumes fell to one-third of the prior best: the V.32 Digicom 9624LE. The unit was highly functional, with ARQ and data compression techniques that had the *potential* for quadrupling throughput (see Section 13.2.4).

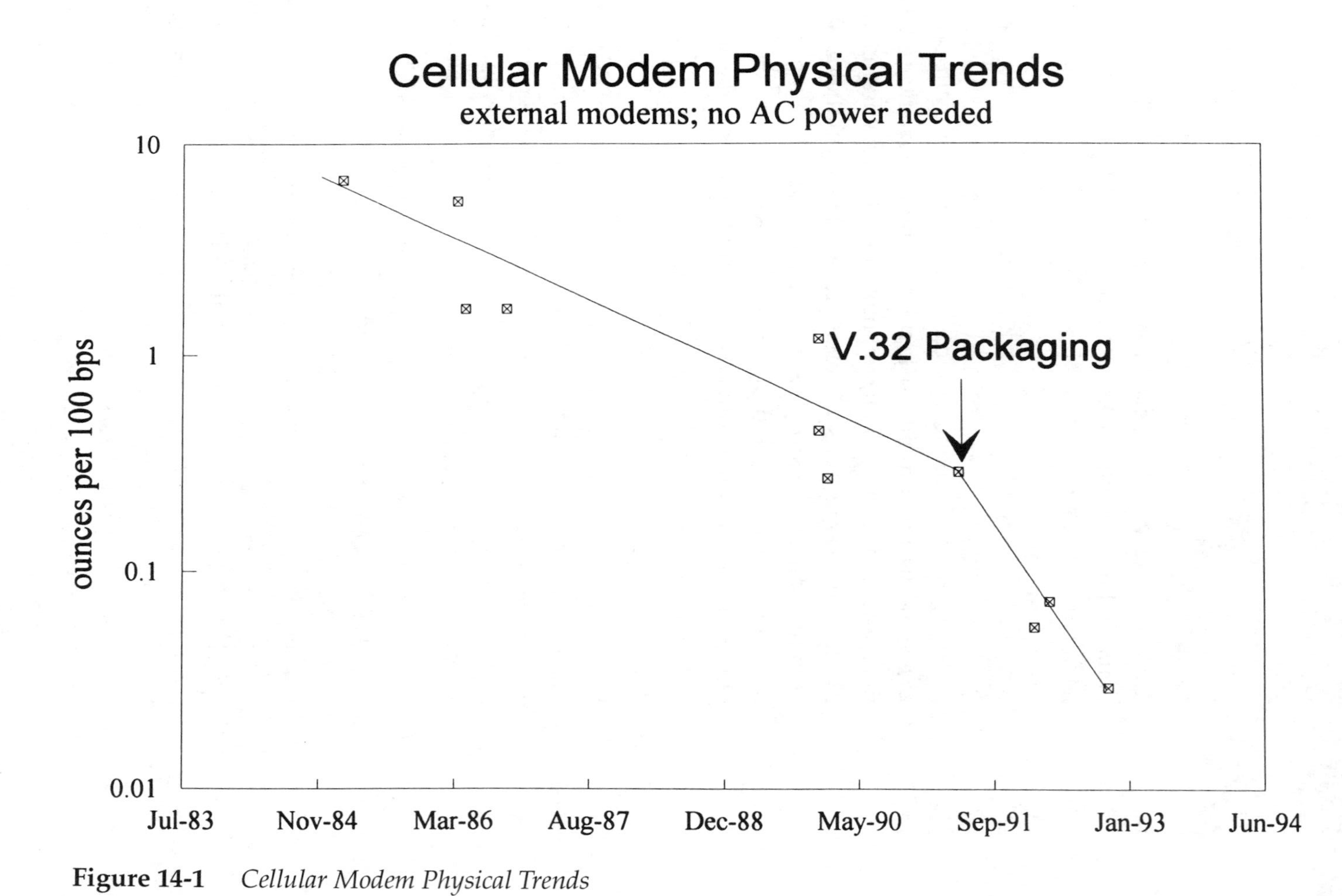

Figure 14-1 *Cellular Modem Physical Trends*

This sharp break signaled the end of the medium-speed standalone modem. Excellent new units such as Microcom's MP1042, in normal times a breakthrough unit, were doomed by the V.32 development. Dozens of vendors[1] began to produce them and the even faster V.32bis pocket modems. Nearly all included fax capability as well. This tight packaging was made possible by the advent of inexpensive chip sets from Rockwell, AT&T, Sierra, and Exar.

List prices fell even faster. In the first quarter of 1985 Spectrum Cellular's Bridge/Span units delivered 300 bps for $595: one bit per second for $1.98. Eight years later Bay Connection's Spectra-Com P1496MX could deliver 14,400 bps (over wirelines) for $299: one bit per second for $.02. That is an astonishing compound decline of 44 percent per year. The Bay Connection price includes 9600-bps fax capability, a 9-volt battery, 9-pin RS-232 cable, phone cord, external power supply, vinyl case, and communications software for both data and fax. Further, direct order retailers sell at well below list.

But the very techniques that produced the pocket modem have now been turned to integration. This will, in turn, doom the pocket modem. There are numerous hybrid integration efforts including placement of Microcom's 1042 electronics, and the MNP protocols, inside Mitsubishi's line of portable phones: the 1500TPK repackaged as the CDL300 (with variants). But the major emphasis is on PCMCIA cards.

Note that effective speeds are rising. AT&T Paradyne claims[2] that its modifications to V.42, called ETC, produce superior performance to that achieved via MNP10. A summary of the claimed advantages includes:

	AT&T Para	Microcom
Average handshake (negotiation) time	23 seconds	57 seconds
Percent success (transfer without disconnect)	97%	68%
Average transmission rate (4K files) : fixed	9600 bps	Unclear
: moving	7200 bps	Unclear

Microcom disputes these findings, especially because their new product, the 4232bis, was not tested. It is obvious, however, that speeds are up and cost/size are down—good news for users.

A point nearly lost is that proprietary forward error-correction modems have lost the battle to open error-detection/retransmit modems. Ironically, in 1992 Spectrum salvaged the interface portion of its business and produces AXCELL/AXSYS interfaces for multiple wireline modems and cellular phones.

14.2.2 Pocket Modems

Table 14-1 is a representative list of current pocket modems with a volume less than 20 cubic inches. The second division separates error detect/retransmit (V.42 only) modems from those also employ forward error correction. As the V.32/V.32bis modems have raw bit rates at least four times faster than the V.22 units—and are as competitively priced (even cheaper in many cases)—the suzerainty of the TCM modems is obvious.

What has begun to differentiate these modems is the software each provides, especially for fax support. Some packages are not compatible with Windows, some are compatible only with specific Windows word processing programs such as Word or AmiPro, some have optical character recognition facilities that permit the user to convert a received fax into editable text, and so on.

Finally, battery life on these little wonders is rarely good—very often well under one hour. More and more often AAA batteries are the best choice. The transmit time improves, sometimes to two hours, and they're easy to find in hotel and airport stores.

	Data bps		Fax bps	List price	Incl soft	G3 C1	Class C2	Error crrct	Error control	Data compression	Extnd. services	Vol. cu in	Wgt ozs	Incl batt
	Basic	Cmprss												
Error Detect/Retransmit:														
Adonics 9624S	2400		9600	$149	Y								6.0	Y
Best Data Smart One Traveler	2400		4800(R)	$199	Y									
Com 1 Voyager MV214	2400		N/A	$390	N	NA	NA		MNP4	MNP5				
Everex Carrier 24/96	2400		9600	$399	Y					MNP5			8.0	Y
Global Village Teleport	2400		9600	$245	Y								8.9	N
Omron Impala 24-96	2400		9600	$399	Y	Y	Y					13.5	6.2	Y
Outbound Pocket Port Fax			9600	$379	Y								3.5	N
Practical Peripherals PM2400PPM	2400		9600	$229	Y							5.06		N
Ventel Pocket Modem 24	2400	9600	9600	$229	Y	?	Y			V.42bis/MNP5		12.4	6.0	Y

Table 14-1 *Pocket Modems:*

	Data bps		Fax bps	List price	Incl soft	G3 C1	Class C2	Error crrct	Error control	Data compression	Extnd. services	Vol. cu in	Wgt ozs	Incl batt
	Basic	Cmprss												
Forward Error Correction Assist:														
Angia PocketStar	14400	57600	14400	$699	Y	Y	N	TCM	V.42/MNP4	V.42bis/MNP5		15.9	8.0	Y
Archtek SmartLink 1414PAV	14400		14400	$349	Y	Y		TCM	V.42/MNP4	V.42bis/MNP5				
Bay Spectra-Com P1496MX	14400		9600	$299	Y	Y		TCM	V.42/MNP4	V.42bis/MNP5			6.5	
E-Tech UFOmate P9696MX	9600	38400	9600	$595	N			TCM	V.42bis/MNP5					
E-Tech UFOmate P1496MX	14400		9600	$299	Y	Y	Y	TCM	V.42/MNP4			19.1	6.35	Y
E-Tech UFOmate P1414MX	14400		14400	$349	Y	Y		TCM	V.42/MNP4	V.42bis/MNP5			6.5	
Hayes Smartmodem Fax144	14400	57600	14400	$599	Y	Y	N	TCM	V.42/MNP4	V.42bis		15.5	8.5	N
Megahertz P296FMV	9600	38400	9600	$399	Y	Y	N	TCM		V.42bis/MNP5		7.82	6.8	Y
Megahertz P2144	14400		14400	$499	Y	Y	N	TCM	V.42/MNP4			7.82	6.3	Y
Microcom Microporte 4232bis	14400	57600	9600	$699	Y	Y	N	TCM	V.42/MNP4	V.42bis/MNP5	MNP10	18.8	10.5	Y

Table 14-1 *Pocket Modems: (continued)*

	Data bps		Fax bps	List price	Incl soft	G3 C1	Class C2	Error crrct	Error control	Data compression	Extnd. services	Vol. cu in	Wgt ozs	Incl batt
	Basic	Cmprss												
Multitech Multimodem MT1432MU	14400		14400	$699	Y	Y		TCM	V.42/MNP4	V.42bis/MNP5			8.5	
Practical Peripherals PM14400FX	14400	57600	14400	$499	Y	Y	Y	TCM	V.42/MNP4	V.42bis/MNP5		11.04	4.2,	
Solectek Transportable	14400		14400	$599	N			TCM	V.42/MNP4			15.0	4.0	
Telebit QBlazer	14400	38400	9600	$299	Y	Y	Y	TCM	V.42/MNP4			13.25	8	
Telebit QBlazer Plus	14400		9600	$599	N	Y		TCM	V.42/MNP4	V.42bis/MNP5			6.5	
US Robotics Worldport	14400		14400	$649	Y	Y		TCM	V.42/MNP4	V.42bis/MNP5		13.2	7.0	Y
US Robotics Worldport	14400		N/A	$599	Y	NA	NA	TCM	V.42/MNP4	V.42bis/MNP5		13.2	7	Y
ZyXEL U1496E Plus	14400		9600	$849				TCM	V.42/MNP4	V.42bis/MNP5				

Table 14-1 *Pocket Modems: (continued)*

14.2.3 Integrated Modems

These units fall into two categories:

1. Custom-shaped units that go under the covers of existing laptop PCs (and, occasionally, cellular phones).

   Originally offered only as extra cost options, in 1992 several manufacturers began to market their laptops with integrated modems as standard equipment. The NCR 3170 is perhaps the best early example with its integrated fax/data "cellular-ready" modem. This low-speed (2400-bps data) unit was made for easy connection to the AT&T 3730 cellular phone.

   After-market integrated modems also exist for laptop upgrades, but not all laptop manufacturers have the volume required for low-cost units. Megahertz has made a market for integrated 2400-bps data/9600-bps fax V.42/V.42bis units of four shapes that fit the following vendor laptops:

Model #	Laptop Vendor
C524FM	Compaq (6 models)
T324FM	Toshiba (11 models/series)
TX324FM	Texas Instruments (4 models/series); Sharp (2 models)
Z324FM	Zenith (7 models/series)

   The list price of these modems span a narrow range from $299 to $329.

2. PCMCIA modems for newer laptops. A representative sample of current offerings can be found in Table 14-2.

PCMCIA Modems:	Data bps		Fax bps	List price	Wrrnty	Type 2.0	Incl soft	AT set	G3 C1	Class C2	Error crrect	Error control	Data comprssn	Extnd'd services	Power (mW)		
	basic	cmprss													oprtnl	stdby	sleep
Error Detect/Retransmit:																	
AMT Starcard 2400fx	2400		9600	$269	5 yrs	Y	Y		Y	Y	n/a	n/a	n/a				
AMT Starcard 2442fx	2400	9600	9600	$299	5 yrs	Y	Y		Y	Y	n/a	V.42	V.42bis				
ComPlus FaxXpress 942	2400	9600	4800	$269		Y					n/a		V.42bis				
CP Plus 2496	2400	9600	9600	$399		Y					n/a	V.42	V.42bis				
Data Race Redi-CARD 2496	2400	9600	9600	$299	5 yrs		fax		Y	N	n/a	V.42/ MNP4	V.42bis/ MNP5				
Dr.Neuhaus Fury Card 2400	2400	9600	2400	$349							n/a	V.42	V.42bis/ MNP5				
DataTrek 2496CFM	2400	9600	9600	$395		Y					n/a	V.42	V.42bis				
E-Tech Research C9624MX	2400	9600	9600	$349		Y					n/a	V.42	V.42bis/ MNP5				
E-Tech Research C9624RX	2400		4800	$299		Y					n/a	n/a	n/a				
EXP ThinFax 9624	2400	9600	9600	$369		Y					n/a	V.42	V.42bis/ MNP5				

Table 14-2 *Packet versus PCA PCMCIA Modems: List Price versus Speed*

PCMCIA Modems:	Data bps		Fax bps	List price	Wrrnty	Type 2.0	Incl soft	AT set	G3 C1	Class C2	Error crrect	Error control	Data comprssn	Extnd'd services	Power (mW)		
	basic	cmprss													oprtnl	stdby	sleep
GVC 9648/24(P)	2400	9600	4800	$269		Y					n/a	V.42	V.42bis/ MNP5				
Hayes Optima 24 Plus Fax96	2400	9600	9600	$399		Y					n/a	V.42	V.42bis/ MNP5				
Intel SatisFax-tion Mobile-Card 96/24	2400		9600	$360		Y					n/a	V.42	V.42bis/ MNP5				
MagicRAM Fax /Modem 332496	2400	9600	4800	$395		Y					n/a	V.42	V.42bis/ MNP5				
Megahertz CC324FM	2400	9600	9600	$379	5 yrs	Y	Y	Y			n/a	V.42/ MNP4	V.42bis/ MNP5		500	450	70
Megahertz XJ124FM	2400	9600	9600	$379	5 yrs	Y	Y	Y			n/a	V.42/ MNP4	V.42bis/ MNP5		500	450	70
New Media PalmModem	2400		4800	$259		N					n/a	n/a	n/a				
New Media InfoBlitz	9600		9600	$299		N					n/a	V.42/ MNP4	V.42bis/ MNP5	MNP10			
Omron Impala 24/96	2400	4800	9600	$399		Y					n/a	V.42	MNP5				
US Robotics Worldport 2496	2400	9600	$349								n/a						

Table 14-2 *Packet versus PCA PCMCIA Modems: List Price versus Speed (continued)*

PCMCIA Modems:	Data bps		Fax bps	List price	Wrrnty	Type 2.0	Incl soft	AT set	G3 C1	Class C2	Error crrect	Error control	Data comprssn	Extnd'd services	Power (mW)		
	basic	cmprss													oprtnl	stdby	sleep
Forward Error Correction Assist:																	
AST	14400	57600	14400			Y			Y	N	TCM	V.42/ MNP4	V.42bis/ MNP5	MNP10			
Angia PCMCIA	14400	57600	14400	$899		Y					TCM	V.42/ MNP4	V.42bis/ MNP5				
AMT StarCard 1442fx	14400	57600	14400			Y					TCM		V.42bis/ MNP5				
AT&T Paradyne KeepIn-Touch:3760	14400	57600	14400			Y	N	Y	Y	Y	TCM	V.42/ MNP4	V.42bis/ MNP5	ETC	1000	15	no
Centennial CC9624FM	9600	38400	9600	$300		Y					TCM	V.42	V.42bis/ MNP5				
ComPlus FaxXpress 1414	14400	57600	14400	$499		Y					TCM		V.42bis/ MNP5				
Data Race RediCARD	14400		14400	$595	5 yrs	Y	Y				TCM						
Dr.Neuhaus Fury Card 14.4	14400	57600	14400	$489		Y					TCM	V.42/ MNP4	V.42bis/ MNP5				
E-Tech Research C1414AX	14400	57600	14400	$499		Y					TCM		V.42bis/ MNP5				

Table 14-2 *Packet versus PCA PCMCIA Modems: List Price versus Speed (continued)*

PCMCIA Modems:	Data bps		Fax bps	List price	Wrrnty	Type 2.0	Incl soft	AT set	G3 C1	Class C2	Error crrect	Error control	Data comprssn	Extnd'd services	Power (mW)		
	basic	cmprss													oprtnl	stdby	sleep
E-Tech Research C1414MX	14400	57600	14400	$449		Y					TCM		V.42bis/ MNP5				
GVC FM144/144(P)	14400	57600	14400	$499		Y					TCM	V.42	V.42bis/ MNP5				
Megahertz CC396FM	9600	38400	9600		5 yrs	Y	Y	Y			TCM	V.42/ MNP4	V.42bis/ MNP5		725	275	77
Megahertz XJ196FM	9600	38400	9600		5 yrs	Y	Y	Y			TCM	V.42/ MNP4	V.42bis/ MNP5		725	275	77
Megahertz CC3144	14400	57600	14400	$599	5 yrs	Y	Y	Y			TCM	V.42/ MNP4	V.42bis/ MNP5		725	275	77
Megahertz XJ1144	14400	57600	14400	$599	5 yrs	Y	Y	Y			TCM	V.42/ MNP4	V.42bis/ MNP5		725	275	77
US Robotics Worldport	14400			$649													
US Robotics HST	16800			$995													

Table 14-2 *Packet versus PCA PCMCIA Modems: List Price versus Speed (continued)*

Increasingly they:

a. Are Group III, type 2 units
b. Achieve 14,400-bps wireline speeds on both data and fax
c. Have V.42 for primary error detection
d. Use V.42bis for data compression

Note the power consumption of the Megahertz modems (one of the few manufacturers to stress this specification). In the 1992 announcement[3] the standby mode was specified at 225 mW, with sleep mode at 8 mW. By April 1993 reality forced the sleep mode up nearly 10 times. Thus, one should watch AT&T Paradyne's specs for similar movement.

Still very much wireline oriented, few manufacturers (Compaq is one exception) support MNP10 for cellular. There is considerable debate, which preceded the AT&T Paradyne ETC claims, about the value of MNP 10 as some "tests indicate that V.42/V.42bis is generally at least as reliable as MNP 10."[4]

One way to distill the dry facts of Tables 14-1 and 14-2 is to compare the low, high, and average price for similar modem types (see Figure 14-2).

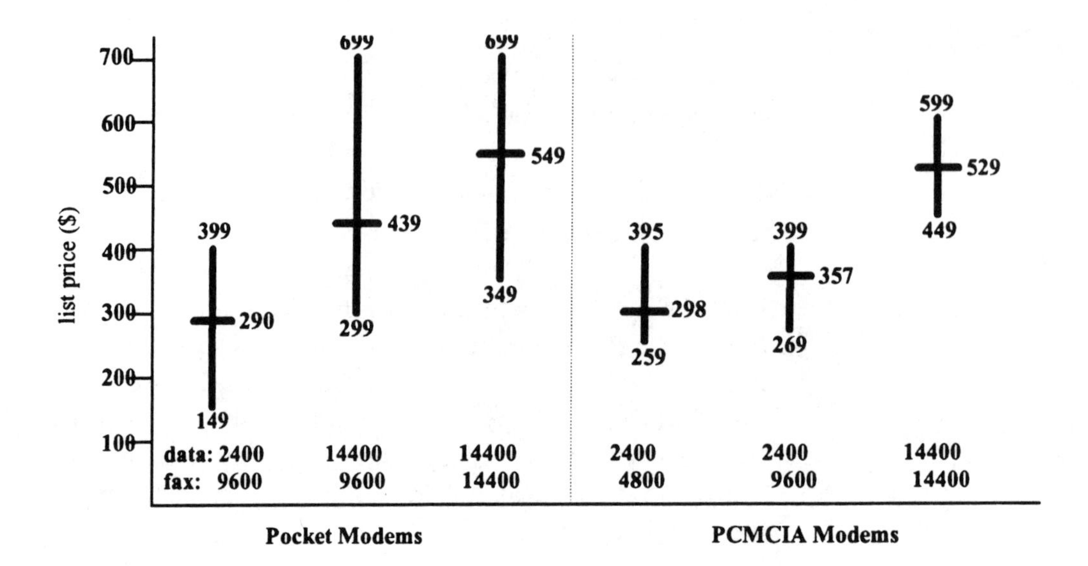

Figure 14-2 *Packet versus PCMCIA Modems: List Price versus Speed*

It is clear that speed has its price; it is also clear that there is no meaningful difference in list price between comparable pocket and PCMCIA modems. As more and more laptops, palmtops, personal assistants, and so on reach the market equipped with PCMCIA slots the pocket modem will pass away.

14.2.4 Modem Interfaces to the Cellular Network

The modems covered in preceding sections were designed with normal RJ-11 telephone connections in mind. A casual examination of most cellular phones (especially handheld units) doesn't yield a convenient RJ-11 jack. If the modem can be plugged into a simple "data link kit" it doesn't take long to notice there is no "send" button on laptop or portable fax. The manual coordination of PC with cellular phone can be an exercise in great manual dexterity.

What is required is a special interface between the RJ-11, wireline world of the modem and the dial-toneless cellular phone (the actual size and location of the interface can vary widely).

Because cellular phone wiring and connectors are far from standardized, the interface(s) are often vendor, even model, specific. The interface must deal with parallel or serial data, positive and negative logic, varieties of coding schemes for pushbuttons and display, balanced or unbalanced audio. And they can be "dumb" or intelligent.

The history of cellular data "jacks" roughly follows this trajectory:

1. In late 1985 NEC introduced its "Computer Interface" (not a particularly catchy name). This $300 external box was a "dumb" interface permitting slow speed (300–1200 bps) modems to plug into its RJ-11 jack. The unit, in turn, plugged into NEC phones. It was simply a conduit to pass information, and required a human being to operate the equipment.
2. In March 1986 Morrison & Dempsey demonstrated the first "intelligent" interface: the AB1X. The key application breakthrough was that the PC could now dial the number. The 30-cubic-inch, 1.5-pound, $400 unit was limited to 1200-bps data transmission, but supported the most common communication software packages as well as cellular phones from 21 different vendors.
3. Later the same year Motorola delivered the "Cellular Connection," a small (~13 cu. in.) external intelligent jack. It connected RJ-11 equipped devices to *most* Motorola cellular phones, providing off-hook dial-tone generation and the standard telephone ring voltage, sensing DTMF, and supporting hook switch flash. Its circuitry permitted 2400-bps transmission but was power hungry.
4. In 1987 Telular obtained an injunction against Morrison & Dempsey for patent infringement. In 1988 the suit was upheld[5] even though Telular "might not have been the first to patent an interface device for use with mobile radio." The AB1X was done; Motorola reached a business agreement for its Cellular Connection.

5. Until 1990 the "intelligent" jack market belonged to Telular, which marketed a series of relatively large (40 cu. in.), relatively expensive (~$500) units. Other vendors concentrated on "dumb" units: Cellabs Datajack, NovAtel's FaxJac, Motorola's Data Passage, and so on.
6. In May 1992 Command Communications began U.S. shipment of the Australian company Intercel's intelligent interface. Called CelCom, it controlled and monitored auto dial/answer from the laptop's parallel port using custom TSR software and V.42 modems. At its initial shipment more than 50 cellular instruments were supported in two general configurations as shown in Figure 14-3.

transportable:

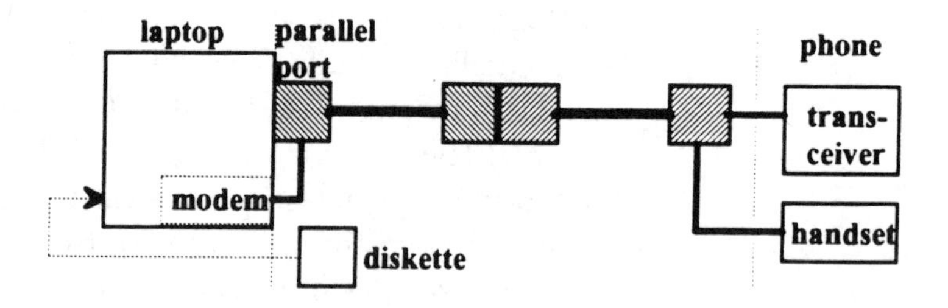

handheld:

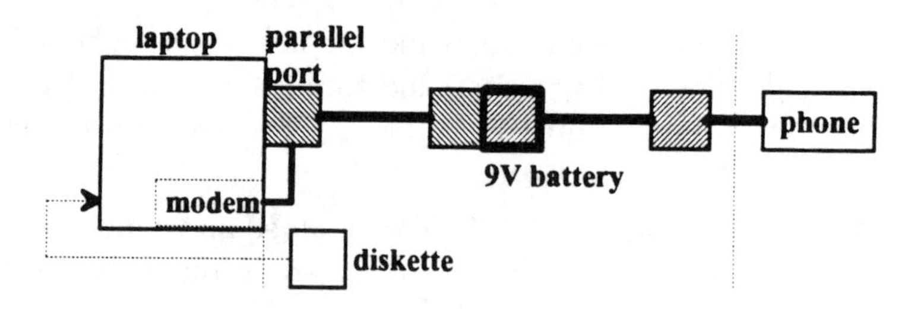

Figure 14-3 *Evolution: Laptop to Cellular Phone Connection*

Later the same year Spectrum introduced its Axcell interface, which connected the RJ-11 jack of a modem to 24 different cellular telephones from 6 different manufacturers. Listed at $395 in advertisements, the Axcell usually can be obtained for less than $300. Spectrum has been uniquely successful, with 60,000 installations by April 1993.[6]

One drawback to the Spectrum approach is the 8.2-cubic-inch, 3.5-ounce adapter unit that physically exists between laptop and cellular instrument. For custom designs the reality of a free-standing box can be masked by splitting the logical functions and distributing them to both ends of the cables. Nokia did this in 1993 with its 121 and PT128 line of handhelds and accompanying PC Card. The handheld phone fits into a "cradle" connected to a multipin PCMCIA data and fax modem card jointly developed with AT&T Paradyne.

The difference between the approaches is shown in Figure 14-4.

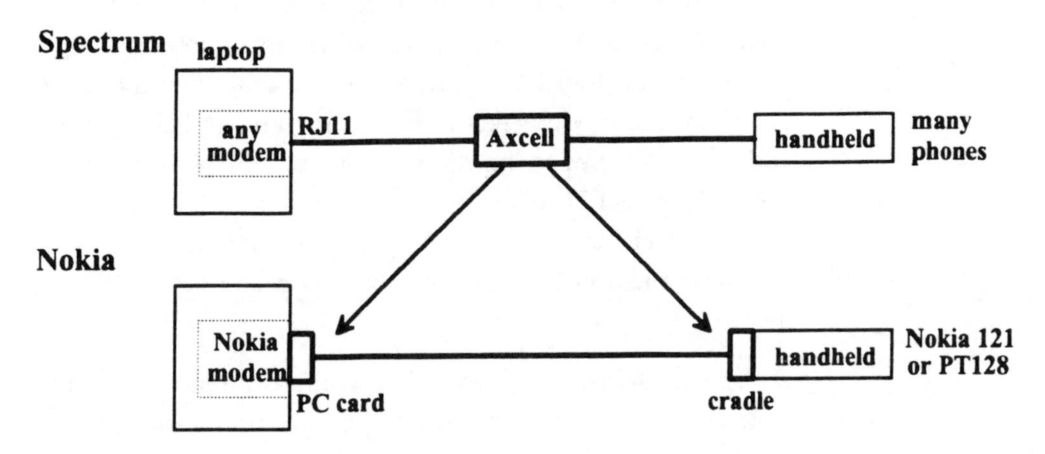

Figure 14-4 *Current Laptop to Cellular Phone Connection*

Other manufacturers, notably Megahertz,[7] have licensed Spectrum technology and will incorporate the functions within their modems, eliminating the need for an adapter in the future.

14.2.5 Cellular Modems with Radios

The use of adapters to connect a wireline modem to a cellular phone is not the only way to send and receive data via the cellular network.

In 1992 Mitsubishi integrated the Microcom 1042 modem under the covers of its 1500MOB series phones. Two of the models—the CDL200 and the CDL300—feature voice handsets. But one model—the CDL100—is simply a modem and cellular transceiver. The modem is V.22, only 2400 bps, and has no fax capability. But it uses the full Microcom MNP10 suite of protocols to improve its cellular readiness.

In 1993 PowerTek, which had been shipping data and fax kits for Motorola Series 3 transceivers for a year, announced its own CMI-3000 cellular data link. Rich with connection options, the $1695 (list) CMI-3000 contains a 3-watt transmitter and a 4000-bps data/fax modem using the Microcom MNP10 correction and V.42bis compression protocols (to add interest to the MNP10 versus ETC dispute, PowerTek claims a 96 percent first-time connect rate with MNP10;[8] AT&T Paradyne claims that *their* tests of MNP10 showed only a 68 percent first-time connect rate.[9])

The basic Mitsubishi/PowerTek configurations differ from Nokia/Spectrum in that a normal connection is from a serial I/O, RS-232 port on a laptop (or other device) directly to the external radio modem—which may or may not have voice capability.

14.3 PACKET SWITCHED MODEMS

14.3.1 Major Trends

Unlike cellular, whose modems are based on wireline developments and usually require some form of rig to get at a radio, packet switched modems are dominated by radio equipped units.

That is not to say that radio-less units do not exist. Some units were modestly successful during the period 1986–1990. Several examples are listed in Table 14-3.

Table 14-3 *Packet Modem Examples*

	Model	Data bps	List price	Ounces	Cu. In.	UBER	FCS
Coded Comm[10]	IQmodem	4800	$950	23.5	58.2	1×10^{-11}	10/90
Dataradio	48MRM	4800	995	24.0	60.0	1×10^{-11}	8/89
Kantronics	RFPM	2400	439		89.7		4/86
RaCoTek	ANM	4800				1×10^{-10}	1990
Western Datacomm	Esteem84	4800	1095	35.2	70.0		1986

With no common industry standard, indeed often proprietary designs; without large volumes to drive prices down; and with a dependency on an external radio (often specific models) that add $500 ± $200 to the price, frequently with addressing restrictions, these modems do not represent mainstream trends. Unless the application demands a specific solution surpassing the limitations of these units—retrofitting data to an existing private voice network is one such need—these units will not be competitive. That circumstance will become less common over time; thus radio-less packet modems are moving toward a historical footnote.

From 1990 to 1993 the mainstream packet switched thrust was external data modems with their own radio packaged inside. Except for a few stray vendors, this category is totally dominated by Ericsson and Motorola, although Gandalf has recently fielded a useful competitor.

Physical trends are not meaningful to plot because the same size/weight modem operates at a wide range of output power and bit rates: the same InfoTAC runs MDC at 4800 bps and RD-LAP at 19,200 bps. Further, that InfoTAC contains radio, cover, power supply, message memory, four-line screen, and eight keys; it is thus a sort of pseudoterminal. It *is* useful to note that in the two-year interval from September 1990 to October 1992, "best-of-breed" radio modems dropped from 3.75 to .83 ounces per 1000 bps with nearly equivalent output power. That is a compound decline of ~52 percent per year.

On June 15, 1993, Motorola announced the major new thrust for packet switched radio modems: a series of PCMCIA type II cards. The two-way versions contain an external antenna/battery compartment that protrudes from the unit housing the radio modem card—instantly dubbed the "tootsie roll." These units will begin to ship during the second half of 1994 on ARDIS and RAM, setting the tone for years of future radio modem developments.

14.3.2 External Radio Modems

Table 14-4 is an edited, but representative, list of the leading external radio modems operating on public packet switched networks. Deliberately excluded are the special-purpose units for private telemetry and SCADA manufactured by vendors such as Motorola (RNet Series), Pacific Crest (PDDR-12), and UDS (DR96).

Vendor	Model	bps	Unit List price	Quantity price	Protocol	Error correction	Error control	Wgt ozs	Vol cu in	Send watts	FCS
Ericsson	C719	8000			Mobitex	8/12 Hamming 20 bit interleave	CRC16, ARQ (go-back-n)	58.0	112.4	6/1	Oct-90
Ericsson-GE	Mobidem	8000	$1,795 $1,395 $775	$1,000 (5000)	Mobitex	8/12 Hamming 20 bit interleave	CRC16, ARQ (go-back-n)	16.0	27.4	2/1	Jul-92 Oct-92 Feb-93
Gandalf		8000			Mobitex	8/12 Hamming 20 bit interleave	CRC16, ARQ (go-back-n)	80.0	102.0		Nov-91
Gandalf	GW200	8000	$999		Mobitex	8/12 Hamming 20 bit interleave	CRC16, ARQ (go-back-n)	16.0	29.9		May-93
Intel	Mobidem AT	8000	$747		Hayes/ Mobitex	8/12 Hamming 20 bit interleave	CRC16, ARQ (go-back-n)	16.0	27.4	2/1	Aug-93
Motorola	InfoTAC	8000	$1,350		Mobitex	8/12 Hamming 20 bit interleave	CRC16, ARQ (go-back-n)	16.0	29.2	3	Oct-92
Motorola	840C11:	4800	$2,500 $1,650		MDC4800	1/2, n=7 16 bit interleave	CRC16, ARQ (stop & wait)	18.0	48.0	4	Sep-90 Feb-92
Motorola	InfoTAC	4800	$1,350 $1,100 $995	 ARDIS ARDIS	MDC4800	1/2, n=7 16 bit interleave	CRC16, ARQ (stop & wait)	16.0	29.2	3	Oct-92 Jun-93 Sep-93
Motorola	InfoTAC	19200	$1350		RD-LAP	3/4 TCM 32 bit interleave	CRC32, ARQ (stop & wait)	16.0	29.2	3	Oct-92
Motorola	MRM420	19200			RD-LAP	3/4 TCM 32 bit interleave	CRC32, ARQ (stop & wait)	41.6	56.1	10	
Motorola	InfoTAC	4800	$1,350		MMP31	45/63 Reed-Solomon	memory ARQ	16	29.2	3	Oct-92

Table 14-4 *Packet Switched Data Radio Modems: External*

Although physical trends of external radio modems are difficult to portray graphically, the price trend lines are more tractable. The steep downward price slope has been spurred, in part, by Motorola's delivery of a competitively priced InfoTAC for RAM. Ericsson reacted with gusto and its Mobidem is now *very* competitively priced. When Intel begins to produce its Mobitex modems, instead of simply repackaging Ericsson units, the price should drop again.

The price slope is shown in Figure 14-5. The compound decline is 33 percent per year for the three-year interval between September 1990 and September 1993.

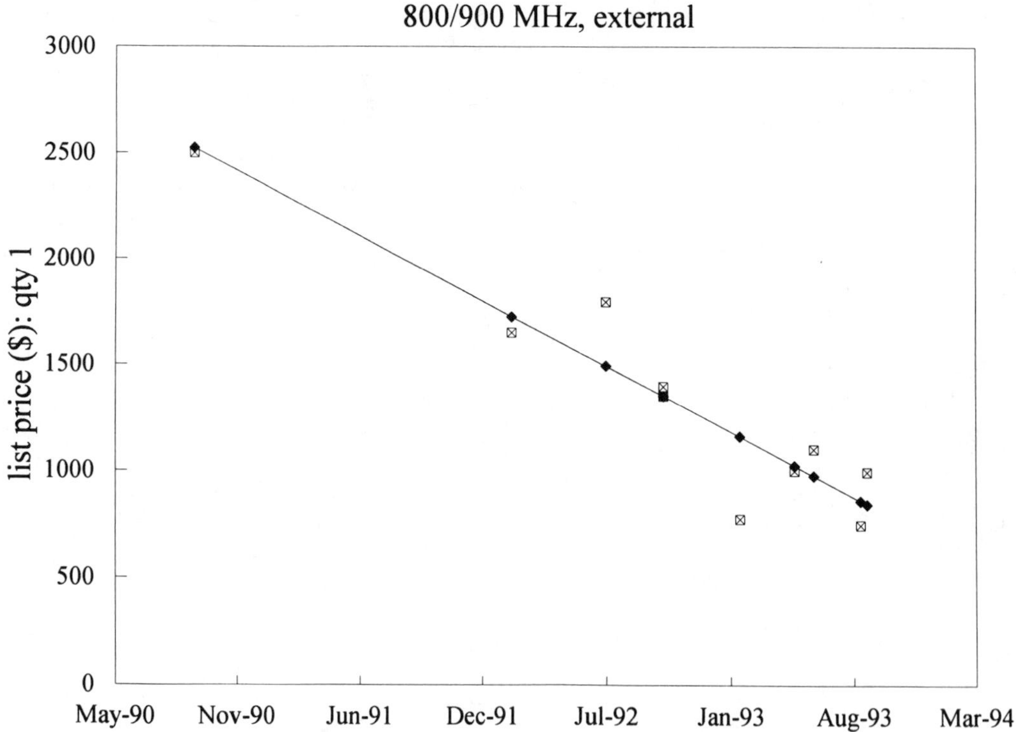

Figure 14-5 *Packet Switched Radio Modem Price History*

Price improvements will be somewhat masked over the next year as list prices hold relatively steady and additional function (such as e-mail support packages) are bundled with the hardware. But the list price barriers are sure to be broken as Intel's Wireless Products organization begins to move modems through its 6,200 retail outlets.[11] Their current plan[12] is for a September 1993 release of a repackaged Mobidem with the AT command set, including either AT&T Easylink or Lotus cc:Mail, for $750.

Even ARDIS, a subsidiary of Motorola, sells the InfoTAC ~20 percent below list.[13] Further, it intends to provide[14] two InfoTAC-based equipment packages in September 1993 for the RadioMail application:

HP package(s)	InfoTAC, leather case, software, plus:	
	HP95	<$1500
	HP100	<$1700
Laptop package	InfoTAC, charger, case, MAC/PC software	<$1000

The lower range of Motorola modem pricing might be divined by examining its 9600-bps, 4-watt RNet series. Although this is a simpler unit than the InfoTAC, its list price is only $690 including communications software packages from Metric Systems.[15]

14.3.2.1 *Vexing Tradeoff #1: Battery Life*

Both InfoTAC and Mobidem have longer battery lives than their circuit switched packet modem counterparts. But that is not to say that users are satisfied, especially because the batteries are both expensive and rechargeable. There is no AAA substitution when the battery gives up the ghost. The typical battery life encountered on the InfoTAC is 3 ± .5 hours[16] (Motorola has an upgrade program in place to improve this time).

In spite of the intense current drain while transmitting, transmit power levels are not as obvious a battery culprit as cellular talk time. In the original KDT-800, the granddaddy of the InfoTAC,

IBM required a battery life of 8 hours, both standby and transmit. Extraordinary weight and size limits were also required.

Motorola prepared a power budget for IBM:

$$\text{Battery Life in Hours} = \frac{400}{38.8 + (.00255 * \text{octets/hour})}$$

If a user's load averaged 400 octets per hour the charge would last 10 hours. If the user averaged 4,000 octets per hour the life dropped to 8.25 hours—the eventual contract level.

At the time (1982) 4,000 octets per hour seemed a very great deal for a 27-character display line. But users quickly accessed PROFS, performing intracompany messaging and storing the long communiqués in the terminal memory. They were then painfully scrolled for viewing (the next Motorola device had a 4×40 screen).

The InfoTAC has moved to control standby power consumption: memory, screen size, and transmit power are down, circuitry is less power hungry. But that circuitry must service synthesizers, not single-channel crystal-cut oscillators. Meanwhile, for certain users message demand has skyrocketed. Employees of the Sheriff of New York have an average airtime service bill of $2,000 per month. Play with tariffs—but their octet per hour total certainly exceeds 9,000.

The message: Obtain a power budget from prospective modem suppliers, match it to your application needs, pick the best modem for your requirements, and buy extra batteries.

14.3.2.2 Vexing Tradeoff #2: The AT Command Set

In 1983 IBM and Motorola struggled to jointly define radio modem product objectives for the proposed public packet switched network. IBM, wary of the development effort required for device applications, favored the Hayes AT command set; Motorola, concerned for the most efficient use of the spectrum, favored native mode. By the time Motorola began the DRN rollout both protocols were present (native mode actually had two variations).

The InfoTAC retains both:

1. Transparent mode: AT command set with industry standard DTE interface for use with off-the-shelf communication packages. Radio channel information, meaningless to (say) ProComm+, is not available in this mode. According to ARDIS[17] there has been little demand for the AT mode.
2. Native mode: a more comprehensive DTE interface for use in RF optimized applications. Motorola's WaveGuide software is used to screen the application developer from too much radio gore.

Ericsson GE faced exactly the same dilemma developing the Mobidem. Upon initial release only native mode (MASC: Mobitex Asynchronous Communications) was supported, designed specifically for sessionless packet switching.

But sales lagged along with application development cycles. The strong hope for e-mail led RAM to development agreements with AT&T Easylink,[18] Lotus (cc:Mail),[19] and Canada's Research-in-Motion[20] (RIM), all of whom favored the AT approach. In March 1993 specially modified Mobidem-ATs were being tested in Canada by Immedia and Mediatel[21] using a RIM package. In June the Mobidem-AT was formally announced in the United States.

Is it the correct solution? Yes! says Ericsson GE's Bill Frezza:[22] "It is going to blow the doors off the wireless industry." No! says RadioMail's President Bill Hipp[23] (and a RAM e-mail provider): "We certainly don't think it is the right use of wireless spectrum."

There is a throughput hit, with a rough yield of 1200 bps.[24] The counter argument:[25] "for transferring long files . . . the . . . AT . . . performance is at least as good as MASC." But the natural riposte to *that* is that long messages belong on circuit switched cellular.

The truth may lie in Hipp's VCR Beta versus VHS example. Beta offered better quality, just as MASC is the better technical solution, but VHS had more movies on tape, just as AT has more big-gun software providers. Stay tuned.

14.3.3 Integrated Radio Modems

Unlike cellular, integrated *radio* modems appeared early in packet switched units. An integrated radio modem was at the core of the KDT800 (FCS: 1983). This smaller-than-a-cigarette-pack unit contained tightly packaged discretes, not VLSI, and had a crystal-cut oscillator, no synthesizer. Still, it fit beneath the covers of a very small handheld unit, transmitted at 4 watts, and had a useful battery life.

Very small integrated radio modems exist in several packet switched devices that operate on ARDIS- or MDC4800-based private systems. Motorola is currently the only supplier of the "piece parts" that form these units.

A first breakthrough was the RPM400i (FCS: 1990) with 3 watts transmit power and 5.7 cubic inches (excluding antenna) space consumption. This unit was adapted by device makers such as IBM's PCRadio (now withdrawn),[26] Lectogram,[27] Poqet's Communicating Computer, and the Telxon PTC860.[28] None of these units was successful.

In late 1991 Motorola announced a successor, the RPM405i.[29] This 3-watt, ¼-inch thick unit required only 1.9 cubic inches of space. The RPM405i was OEM priced for vendor integration at $500–1,000, depending upon quantities.[30] At 10,000 units the price was $800.[31] It was quickly adopted by Psion PLC's HC,[32] Toshiba's 486-based laptops,[33] and has been released for use on ARDIS as the PDT220. Announced initially at 4800 bps, it has become the core for both the InfoTAC and the MRM420 19,200-bps modems.

Toshiba is now using the 415i in laptops, where it displaces a diskette drive[34]; the 435i variation is scheduled for Mobitex protocols in 1993.[35]

Thus, the OEM price of a high-speed modem—with radio—was theoretically competitive with equivalent prices of wire-line/cellular PCMCIA counterparts in 1H93:

	Price	Transmit Rate	User Rate	User $/bps	Radio
Motorola 405i	$800	19,200	14,400	$.056	Yes
Megahertz (OEM qty)	489	16,800	14,400	.034	
Spectrum Axcell	300	—	—	.021	
	$789			$.055	No

especially because they were self-contained and the PCMCIA modem requires a cable to a radio.

But vendor markups were steep for the necessary physical packaging of the 405i into cradles or docking stations. Further, PCMCIA prices are dropping; the Megahertz *unit* price was down to $499 by 2H93. And the PCMCIA-enabled PC forecasts cannot be ignored: One estimate is that there will be at least one PCMCIA slot in all handheld computers and 70 percent of all notebook computers by 1995.[36] The need for Motorola to counter with its own PCMCIA cards is inescapable.

But the engineering task is formidable. The modems must be protected from PC radiation. In turn they must be able to radiate in ways that do not harm their host. Laptops with PCMCIA slots in front may not be able to use the radio modem (besides, who wants to think about what that little guy is doing radiating into your face?). The power output has been reduced to ~1 watt, forcing an external antenna to improve the effective radiated power (ERP). Battery life, as always, will be a problem in such a small device.

Current wisdom suggests that Motorola will begin testing PCMCIA radio modems with ARDIS in late 1993 with customer delivery for both ARDIS and RAM occurring in 2Q94. Toshiba appears to have a plan in place to install these units in its 4500/4600 laptop follow-ons.[37] A CDPD version is to follow in 1995. That unit has even graver engineering challenges because it must be truly full duplex to be of interest. But the pressure is on. Cincinnati Microwave claims it will have a $300 data radio modem—possibly an external unit—ready for 1Q94 CDPD customer ship.

But modem developments have a history of not meeting their schedules. Ericsson's Mobidem was plagued with microcode bugs (flushed by RadioMail) when it attempted to enter sleep mode. Motorola is plagued by roaming bugs (also flushed by RadioMail) on its InfoTAC; its dual-protocol modems were a year late for ARDIS in Washington, D.C.

The Motorola PCMCIA radio modem may be late (as may Cincinnati Microwave) but the trend is clear: fully integrated units at competitive prices.

14.4 MULTIFUNCTION RADIOS, WITH MODEMS

Promotional information[38] accompanying the release of the multistage CDPD specifications state: "The cellular networks supporting the CDPD technology will offer a broad range of wireless data and voice applications for mobile workers by enabling *one portable computing device* to offer all of their office needs—phone, e-mail, fax and computing [italics mine]."

How can this be possible? Logically, CDPD shares only spectrum with the cellular network. And one has a hard time speaking into a keyboard! Thus this somewhat puffy reference must be to a composite workstation: a PC with handset, a 3-stage modem (V.32bis for data over cellular, V.17 for image fax over cellular, and the GMSK/CDPD data protocol), and a shared radio.

If the radio is shared then it can be in, say, voice mode using the handset over cellular. Or it can be in packet switched data mode via CDPD. It can't be doing both tasks simultaneously.

Because cellular-class radios are tumbling in price, the cost advantages do not appear to be overwhelming. The flip side of the cost observation is that the workstation might hold two radios, one for voice, one for packet switched data. But two 800 MHz radios transmitting within an inch (or so) of each other, at possibly 3 watts each, poses interesting technical problems. A more likely supposition is that voice and data will not be transmitted *simultaneously* until one of the two applications moves to an alternative technology, with CDMA being a likely candidate.

If a single radio can be shared in an AMPS/CDPD environment, can that same radio be shared in an AMPS/ARDIS or AMPS/RAM environment? Possibly, with the following caveats:

1. There are frequency band differences. ARDIS is closest to AMPS cellular because it operates in 806–824 MHz. RAM, however, operates at 900 MHz.
2. There are channel spacing differences. Cellular (and thus CDPD) has 30 KHz in which to play; ARDIS has only 25 KHZ; RAM, only 12.5 KHz.
3. ARDIS has adjacent channel considerations; it operates on any channel it can procure in a given metro area, not on a predictable, contiguous group.

Further, if a full-duplex radio becomes available to ARDIS the protocol must be adjusted, and base stations upgraded, to exploit the new capability. This is quite possible—and would be a useful inbound performance boost—but would take time.

REFERENCES

[1] *PC Computing* report on high speed fax modems included tests of 48 different modems, June 1993, p. 288.
[2] Vector development test results as reported by AT&T Paradyne, *Mobile Data Report*, 7-5-93.
[3] Megahertz PCMCIA Notebook FAX/Modems Technical Specification.
[4] Spencer Kirk, Megahertz President, *Mobile Data Report*, 9-28-92.
[5] Judge J. Spencer Letts, U.S. District Court Judge for Central California, Santa Ana, as reported in *Cellular Sales & Marketing*, November 1988.
[6] Peter Caseta, Sr., President, Spectrum Information Technologies, *Mobile Office*, April 1993, p. 34.
[7] Spencer Kirk, President, Megahertz, *Mobile Data Report*, 9-28-92.
[8] PowerTek CMI-3000 promotional brochure, p. 7.
[9] Vector development test results; *Mobile Data Report*, 7-5-93.
[10] Now merged with Decom Systems, *En Route Technology*, 8-30-93.
[11] *Microcell Report*, March 1993.
[12] *Telocator Bulletin*, 6-18-93.
[13] *Mobile Data Report*, 6-7-93; confirmed at the ARDIS Lexington Conference, 7-27-93.
[14] ARDIS Lexington Conference, RadioMail pricing.
[15] Motorola Paging & Wireless Data Group price through 10-29-93.
[16] *Mobile Data Report*, 6-7-93.
[17] Crystal Cooley, ARDIS Technical Specialist; *Mobile Data Report*, 3-15-93.
[18] *Electronic Messaging*, 7-8-92.
[19] *Electronic Messaging*, 11-11-92.
[20] *Enhanced Services Outlook*, December 1991.
[21] Michael Ham, Cantel Director of Sales & Marketing; *Mobile Data Report*, 3-15-93.
[22] *Mobile Data Report*, 5-24-93.
[23] *Mobile Data Report*, 6-21-93.
[24] Dan-Haken Davall, Director of Ericsson's Canadian Mobile Data Group, *Mobile Data Report*, 5-24-93; Glenn Kaufman, Lotus Wireless Product Manager, *Mobile Data Report*, 6-21-93.
[25] Martin Levetin, RAM Senior Vice-President for wireless messaging; *Mobile Data Report*, 6-21-93.
[26] *En Route Technology*, 8-30-93.
[27] *Mobile Data Report*, 9-9-91.
[28] *Industrial Communications*, 12-13-91.
[29] *Advanced Wireless Communications*, 12-11-91.

[30] *Communications Week*, 12-16-91.

[31] Douglas Fraser, Motorola MDD Product Manager, as reported in *Data Communications*, February 1992.

[32] *Mobile Data Report*, 12-16-91.

[33] *Mobile Data Report*, 8-16-93.

[34] *Edge On & About AT&T*, 6-14-93.

[35] *Mobile Data Report*, 8-2-93.

[36] InfoCorp as reported by Andrew Seybold's *Outlook on Mobile Computing*, vol. 1, no. 6, p. 15.

[37] *Mobile Data Report*, 8-16-93.

[38] Waggener Edstrom, CDPD Release 0.9 Announcement, 5-11-93.

PART V

PUTTING IT TOGETHER

CHAPTER

15

USER APPLICATIONS

15.1 VERTICAL VERSUS HORIZONTAL

The trade press is rife with conventional wisdom reports on vertical versus horizontal markets. ARDIS is "known" to focus only on the verticals because of its "heritage and infrastructure limitations,"[1] or its "genetic" differences.[2] RAM, of course, is thought to focus only on horizontals.

What do these buzzwords mean?

To most people a "vertical" is application-specific: perhaps parts order entry from a mobile device transmitted upward to the same company's host computer for processing. The applications are highly tailored because field service firms, for example, often see the intricacies of their particular part stocking response as a competitive edge. The applications are not generic and do not transfer well from, say, high-tech medical/computer field service to truck-repair field service—or to public safety or field sales, for that matter.

A "horizontal" cuts across market segmentation boundaries with generic tasks: e-mail or credit card authorization, for example.

It is clear that RAM currently does have more focus on horizontals than ARDIS. Its rich array of business alliances, especially in e-mail, clearly testify to that drive. It has not yet made a successful business on the horizontal market. It *is* successful with selected verticals (RAM is not exclusively horizontal) such as Chicago Parking Authority and Conrail.[3]

ARDIS began with a handheld device with a strong field service bias. The only radio modem alternative for the laptop user (a market only then beginning to explode) was both external and expensive. E-mail was reasonably well understood. About one-third of IBM traffic, which accounted for 92 percent of all ARDIS devices in 1990, was messaging (nonmailbox, int*ra*company). But IBM could not calculate displaceable costs based on messaging and so the horizontal drive was stifled. So, too, was the attempt to provide "fax servers." Customers clearly wanted true image fax, not cost justifiable on packet switched networks, and used cellular instead. ARDIS began to focus more and more on vertical markets where it could make money.

But the focus changes constantly. ARDIS application profiles revealed that two-way messaging was "critical to the functionality or economics" of 20 out of 22 application profiles.[4] E-mail (mailbox, int*er*company) was not critical to the bread-and-butter field service application, but began to appear more often as a critical requirement in field and insurance sales, financial services, lawyers, accountants, consultants, executives/managers, and so on. (Incidentally, many of these applications also require voice.) The requirement was not lost on coparent Motorola, which invested $2 million for a minority share in RadioMail.[5] ARDIS is back on the horizontal track with an e-mail availability date of September 1993, delayed by roaming bugs in InfoTAC microcode.[6]

15.2 VERTICAL APPLICATION EXAMPLES: FIELD SERVICE

15.2.1 IBM's DCS

IBM's Field Service dispatch history stretches back decades to pager-based systems (until about 1984 IBM had the largest private paging system in the United States). Split into two service divisions in the 1970s, each division pursued alternatives to pure paging.

The first change came with voice pagers. The Field Engineering Division (FE), after measuring thousands of dispatch calls, developed a dependable time profile as shown in Figure 15-1.

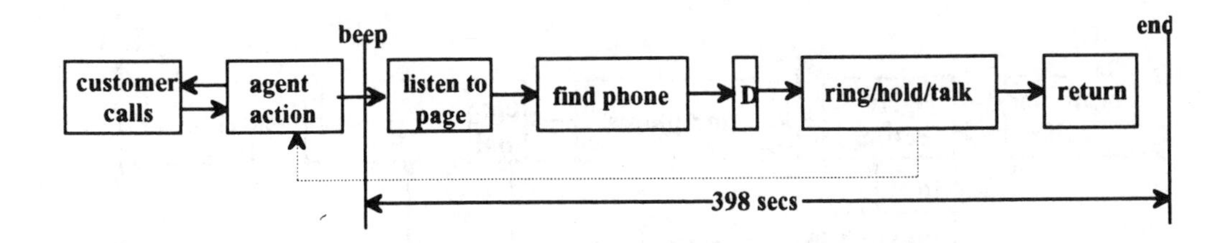

Figure 15-1 *"Beep-to-Complete": Field Service Dispatch via Alpha Paging*

1. The IBM customers called an 800 number that connected them to an agent at a regional dispatching center.
2. Using custom application packages that identified customer, maintenance contract status, equipment location, primary/secondary field service personnel, etc.—and listening to the customer's problem—the agent made a voice page to the right service person (primary could be on vacation, already on a call, or otherwise unavailable).
3. When the service person got the beep an average of 72 seconds was required to listen to the voice page.

4. The service person then required, on average, 120 seconds to find a phone (usually a customer phone, which was not always appreciated). Dial time required an additional 10 seconds.
5. The ring/hold/talk time required, on average, another 136 seconds. If the talking was to the dispatch agent in order to "sell the call" (reassign it because the service person was already engaged) there was inevitable time lost asking how the Cubs did last night.
6. On average, the service person was back at work in 60 additional seconds.

Total "beep-to-complete" time was thus 72 + 120 + 10 + 136 + 60 = 398 seconds.

Meanwhile the Customer Service Division (CS) had been experimenting with portable terminals. It had developed a small unit that altered the dispatch flow in the following way, as shown in Figure 15-2.

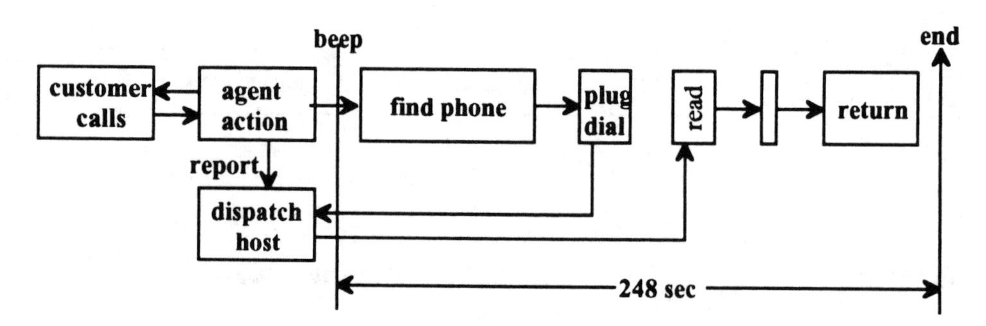

Figure 15-2 *"Beep-to-Complete": Field Service Dispatch, Page plus Terminal*

1. The voice pager was replaced with a simple tone pager (IBM tends to be antivoice).
2. When the customer called, the agent filled in the trouble report online to the regional dispatch host and beeped the service person.
3. The service person took 120 seconds to find a phone.

4. Picking up the unit (3 seconds), plugging it into an RJ-11 jack (15 seconds), dialing the number (same 10 seconds), and handshaking with the dispatch host (2 seconds) came next. Subtotal: 30 seconds.
5. The dispatch host then responded with the trouble report previously entered by the agent. The read/response time: 25 seconds.
6. Next the terminal was unplugged (10 seconds) and put down (3 seconds).
7. The service person was back at work in the same 60 seconds.

Total "beep-to-complete" time was thus 120 + 30 + 25 + 13 + 60 = 248 seconds.

It took no rocket scientist to deduce that most of the 136 seconds spent in agent ring/hold/talk time vanished with approach #2. Call it two minutes. It was clear that taking the dispatch via a data message would radically reduce the number of dispatch agents, their tubes, desks, even offices (some regional centers were closed in the consolidation).

The next step was to go after the service rep time consumed by telephone search and dial. The only answer was wireless; several inhouse prototype units were built and tested. They led to interesting business case discoveries:

1. When the device beep occurred the dispatch message was already present. With a small unit the service rep could pick up the device in 3 seconds.
2. Reading the message and keying the response (say, to accept or sell the call) took 14 seconds.
3. Putting the device down took 3 additional seconds.

Now the beep-to-complete cycle had contracted to 20 seconds. Automated dispatch was getting interesting. Further, the system absolutely knew that the message had been delivered and could begin seeking alternative service personnel quickly if it had not been (there was always a time uncertainty with paging).

Further, it was clear that two other payoff areas existed:

1. Parts order entry: This embraced not just ordering a part but also checking on the delivery status and changing priorities. It was later expanded so the field service rep could be told that the part sought was in the city warehouse five blocks away.
2. Incident reporting: This application caused the failure source, its part #, and symptoms to be sent to a common database computer. Summary reports were then sent to the responsible engineering group for early warning use.

and other areas could not be justified:

1. Messaging, as noted in Section 15.1.
2. The use of the portable wireless terminal as a plug-in diagnostic aid.

Building on years of steady application development, and knowing which applications paid off and which did not, and knowing further how rapidly the new system must come up and counting on resale of the paging system (which did happen), the DCS contract was signed in 1981.

A classic vertical system, it took on the following form in the network as shown in Figure 15-3.

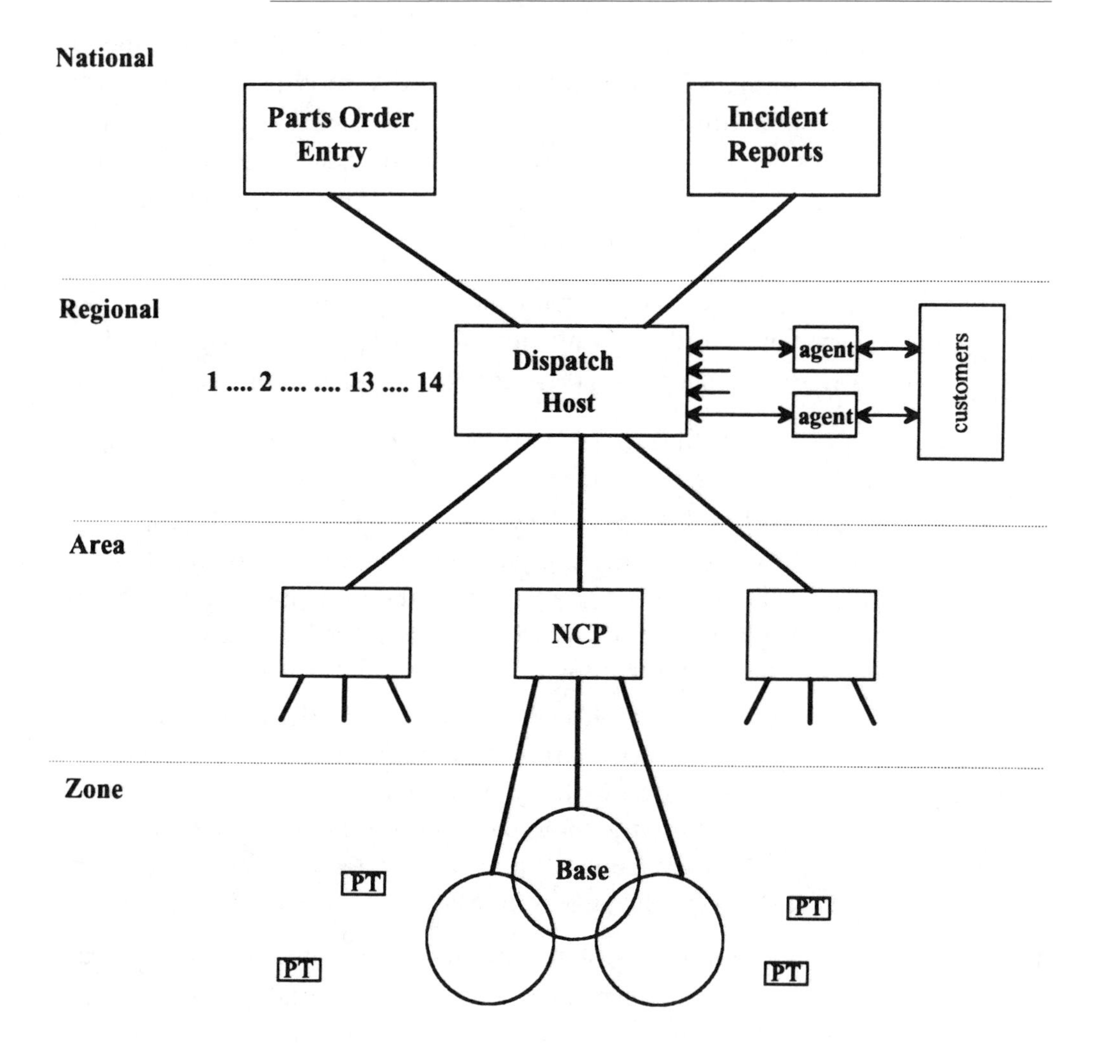

Figure 15-3 *Vertical Application: Field Service*

The customer calls in the trouble report to a regional dispatching center (once there were 14; the productivity gains and the "big blues" have reduced that total). The dispatch agent posts the problem to the regional hosts.

The regional host "pages" the field service person of choice. That is, it sends out the briefest "are you there?" inquiry to the last known location of the service rep. The page is highly likely to get through because of its short length. It also requires no ACK, an application requirement that modifies how the radio system functions. IBM was terribly concerned that a 4-watt automatic acknowledgment kicked off from deep within a large processor could cause customer mischief. The first contact is ACK-less.

The field rep receiving the inquiry is required to move to a safe transmit area, if not already there, and respond with an "I am here" message. This brief, manual ACK does three key things:

1. It lets the *application* system know that the selected service rep is available.
2. It lets the *data radio* system know the received signal strength from the device. Note that it would have been theoretically possible for the device to also report the signal strength it had received from the base station. This was not implemented.
3. It starts a "time slice" running in the device. As long as the user variable time slice (typically 30–45 seconds) is running automatic ACKs will be invoked. It is the user's responsibility not to move back into a danger zone while interactive communication is underway. Each action by the user during this interval rekeys the time slice.

The application system now responds with the full dispatch text to the radio system. The radio system selects the best route, aided by signal strength indications from multiple base stations that have received the "I am here" message. It thus delivers the full text with high (90 percent) probability of success on the first attempt. Retries are the responsibility of the radio system; the application is unaware of any underlying messiness.

A variety of things can happen next. The most common is that the service rep takes the call. The time is posted back to the application and the customer's trouble call updated online. If the problem is complex or unfamiliar, there is often peer-to-peer messaging as the rep asks buddies if they've ever seen this problem. If it's *really* bad, or the customer is livid, the rep may contact the field service manager for assistance.

When the trouble is corrected the rep closes out the incident and possibly back orders parts. These message types arrive at the Regional Dispatch Center first; it realizes they are intended for another system and forwards them over the interconnected SNA links to the proper destination.

Built into the application, not the radio system, are contact scenarios. If rep "A" has primary responsibility for the customer, and the radio system is unable to get a response in four tries, that fact is reported to the application. A broader search pattern can be requested by the application (very rare) and the radio system will range outside home areas (it *can* broadcast if it has to).

Typically the application logic will decide that:

1. This is not a burning problem (say, an annoying flicker on a PC) and it will initiate a request for rep "A" every 15–30 minutes. People often like to eat in peace; the rep may have turned off the terminal. After an hour the application program may revise its strategy to step 2.
2. This problem can't wait. The radio system is instructed to find rep "B."

Note that an underlying assumption of the system design is that the service rep is not wide ranging. If someone does jump in the subway at Grand Central, cutting off communication, that person is expected to signal that he or she is once again reachable upon surfacing at the Battery.

Ironically, the initial infrastructure design seemed to indicate that the system was intended to let the field service rep roam the nation. IBM secured a common nationwide channel from the FCC (with some exceptions in areas abutting the Mexican and Canadian borders). But the goal was far more pedestrian. The handheld device's weight/size/battery-life targets were so stringent that Motorola was unable to include a synthesizer.

The KDT terminals went with crystal-cut oscillators. Because IBM was concerned with spare stocking, matching devices to numerous possible city frequencies would have been a logistical nightmare; ergo, a common frequency.

Coverage requirements were also plotted with great precision. The nation was broken up into two-kilometer zones and service rep populations were projected for each zone. There are enormous population differences between Manhattan, New York, and Manhattan, Kansas. It was thought that a significant number of service reps would never receive radio coverage. Thus the KDT also had a 300-bps integrated modem standard, as it might be useful in a base-station down situation. The KDT was priced with and without radio transceiver.

As the rollout proceeded, and as the base-station reliability was measured, the decision was made to ship all KDTs with radio transceiver in place if only to minimize spare parts stocking problems. In fact, radio coverage became ubiquitous, extending to the business areas of Alaska and the Caribbean.

Field trials were held with the marketing divisions in an attempt to establish a business case for them (under the terms of the IBM purchase agreement there were going to be plenty of devices available). Sales personnel loved to be notified when a customer's mainframe had been down for two hours with the problem not yet diagnosed. They were able to speed over and reassure the customer that they were on top of the situation. But a price value could not be placed on such information and the field trials were abandoned.

New, cost-justifiable applications could be envisioned. However, they usually did not fit in the memory constrained KDT800, and they were always hard to develop. There was also the usual risk of committing them to read-only memory when still unstable. As IBM begins to restructure into independent companies, Pennant Systems (printers) has begun to move away from the KDT to laptop based units that employ integrated radio modems. The day of the special-purpose handheld radio unit is probably over, though Husky, Itronix, Melard, Telxon, and others would certainly dispute that conclusion.

15.2.2 Pitney-Bowes AIM

Roughly five years after IBM's pioneering field service work, Pitney-Bowes began to plow similar ground. Like IBM, it had developed (1985) a mainframe-based field service system called ACESS: Automated Customer Support System. This was essentially a telephone-based system. Service reps called in at the end of the day to obtain the first call for the following day. When they called in to report the call complete they received their next dispatch. Only special field service reps were given pagers; thus the managers could not initiate calls to the service rep. Further, Pitney-Bowes knew that its customers disliked field service reps' use of their phones.

In January 1990 ARDIS was announced; the two companies were at work by February. The agonizing application specification work, cost justification, and coverage tests consumed a year, without a contract in place.

Large-scale implementation took nearly another year. Some mistakes were made, especially in terms of message volume and length. This problem was compensated for by doubling the size of the front-end processors and a great deal of system optimization work at the handheld device (the KDT840, successor to the KDT800), including data compression and transmission of only changed data (see Table 15-1).

Table 15-1 *KDT800 versus KDT840*

	KDT800	**KDT840**
Volume	46.5 cu. in.	46 cu. in.
Weight	31.0 oz.	30.0 oz.
Environment	Rugged	Rugged
Memory:		
Capacity	240 KB	512 KB
RAM	80 KB	320 KB
EPROM	160 KB	192 KB
I/O port	Custom	RS-232 (TTL)
Application creation	Assembler	BASIC; C
Display:		
Characters	2 × 27	4 × 40
Backlit	No	Yes
Keyboard	59	49

There were also registration problems, which were satisfactorily patched for Pitney-Bowes but forced ARDIS to deal with its lack of automatic roaming.

By the beginning of 1993 more than 2,800 users were on the system. The payoff was clear:[7]

1. 18 dispatch centers consolidated to 6.
2. Dispatch agents reduced by 60 percent.
3. 5 percent increase in rep productivity (there were reductions in the number of reps).

while Pitney-Bowes machine inventory grew 4 percent.

There were also problems. In spite of the fact that the KDT840 has considerably more memory than its predecessor, it's not enough. Further, the lack of a synthesizer hurt Pitney-Bowes, which had to stock spares with 10 different frequency allocations. The plan is to abandon the KDT840, beginning in 1994, and use Motorola's integrated 405i radio modem in DOS-based laptops.

Radio coverage patterns differ from IBM's so that dead spots exist. Further, Pitney-Bowes wants multiple public airtime service providers to drive down airtime costs. ARDIS has committed to work on a gateway processor that will permit Pitney-Bowes to reach other networks, including satellite offerings.

As happened at IBM, experimentation is underway with the field sales personnel. But these 150 test users have IBM notebook computers, which improves flexibility. It is not yet clear that the sales applications can be cost justified.

15.2.3 The Generic Approach

The overriding message of the DCS and Pitney-Bowes examples is that vertical applications demand enormous planning, long development cycles, an understanding of cost benefits, and a tailoring of the radio system and devices to meet the application needs.

It is no surprise that ARDIS chose to concentrate on field service as an initial application. They understood the problems, many of the staff having bled through DCS, and had a unique device with integrated radio modem (KDT840).

Still, a typical customer sales cycle (this is EDS) required more than two years (see Table 15-2).[8]

Table 15-2 *Typical Sales Cycle: Vertical Application*

July 1991	Decision-maker call
August 1991	ARDIS application study
September 1991	Proposal made; starter kit contract signed
Oct.–Dec. 1991	Pilot
January 1992	Evaluate pilot results
Feb.–Apr. 1992	Application requirements (re)defined;
	Customer reference visits;
	Lexington/Lincolnshire executive briefings;
	devices selected
May 1992	Requirements complete; application work started
July 1992	Services contract signed
Aug.–Nov. 1992	System develpment & test; 11/15/92: 25 unit test
Dec. 1992	Application/hardware delay production rollout
Jan.–Feb. 1993	Rollout begins
Mar. 1993	Hardware reevaluation; suspend rollout
Apr.–Aug. 1993	Rollout resumes and concludes (700 users)

This enormous up-front investment in each customer, accurately foreseen by the IBM participants in the first IBM/Motorola exploration of the airtime service business, is akin to having a business selling clarinets. Not a bad business, but not a big one, either, because clarinets are too hard to learn to play (apologies to Paul Carroll).[9] ARDIS resources were strained by the requirement to provide the end-to-end implementation services shown in Figure 15-4.

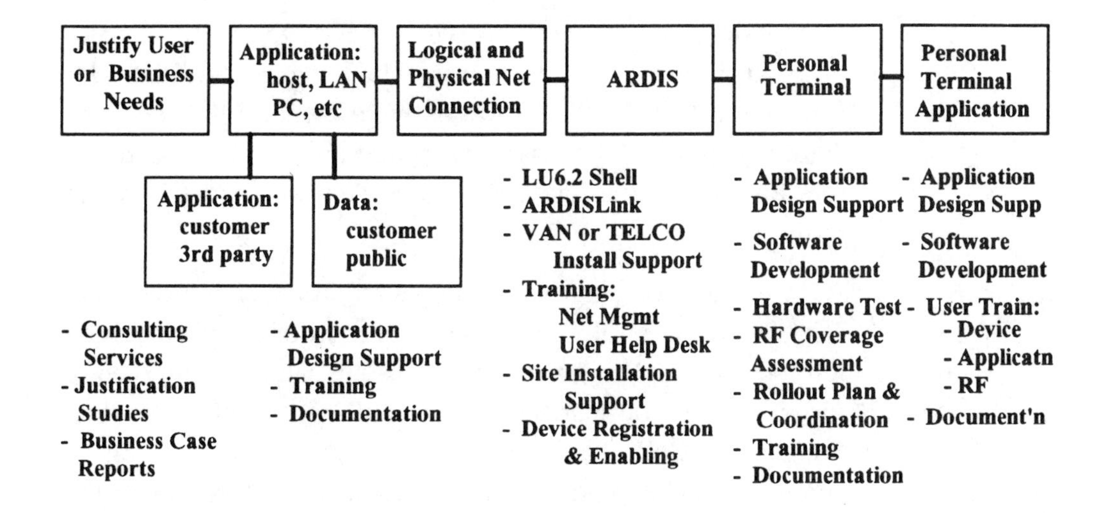

Figure 15-4 *ARDIS End-to-End Implementation Services*

It seemed clear that there would be no way to make the necessary investment to reach the small (<100 users) vertical application. Inability to reach the low-volume user has long been seen as a crippling obstacle to market expansion. In March 1987 a detailed examination of voice SMR user distribution in Manhattan was made by JFD Associates.[10] One summary of size categories is given in Table 15-3.

Table 15-3 *User Size Distributions:Manhattan*

User Size	# Users	% of Total	Cum. from Largest	# Radio	% of Total	Cum. from Largest	Radios per User
1	498	23.0%	100.0%	481	3.0%	100.0%	1.0
2–3	416	19.2%	77.0%	1055	6.6%	97.0%	2.5
4–7	648	29.9%	57.8%	3484	21.9%	90.4%	5.4
8–15	374	17.3%	27.9%	3904	24.5%	68.5%	10.4
16–31	166	7.7%	10.6%	3582	22.5%	44.0%	21.6
32–63	51	2.4%	3.0%	2166	13.6%	21.6%	42.5
64–127	10	0.5%	0.6%	803	5.0%	8.0%	80.3
	2,166			15,934			

In that long-ago time there were 15,934 voice SMR radios licensed in New York City, many of them logical candidates for conversion to data. These radios were owned by 2,166 businesses (courier and messenger, plumbing, electrical contractors, towing services), a discouragingly low average of 7.4 voice radios per user.

A ray of hope for a company such as ARDIS (which did not yet exist) was that just three users (0.1 percent of total) accounted for 468 radios (2.9 percent of total). The sales and marketing focus could be on just three accounts—two of which (Yellow Freight and DHL) happened to be nationwide. But calling on enough users to account for 21.6 percent of the device potential meant 64 sales efforts; 51 of those calls were on prospects averaging only 42.5 devices.

This problem led ARDIS to the development of ServiceXpress, a bundled flat monthly fee offering created by a four-way alliance:

1. ARDIS
2. SonicAir (courier and warehousing logistics provider)
3. Service Systems International (field service applications software)
4. Business Partners Solutions (connectivity solutions)

As expected, the emphasis is on field service applications. Parts information is held on SonicAir's inventory management system and delivered by its couriers. There is no up-front capital expense because the system is "pay as you go" ($250/month × 31 months) for airtime, subscriber units, dispatch terminals, maintenance and repair, and key applications: Call Handling (dispatch, messaging, status/inquiry, incident report) and Parts Logistics (inquiry, requisition, receipt, transfer).

ARDIS was delighted to have a six-technician company sign up by telephone. A more typical response was the field service operation of Great Western Bank:

November 1992	ServiceXpress announced; Great Western signs
December 1992	Contracts with BPS, DCS, SonicAir, SSI
January 1993	Training: ARDIS, BPS, SSI
February 1993	Training: DCS, SonicAir
March 1993	GW field test installation (2 days)
	GW connection to existing application (15 days)
	Production

which shortened a two-year cycle to five months. The success at Great Western led to reference sales at Colonial Business Equipment, Siemens Medical, and CompuComm in April, and to the qualification of nine more prospects thereafter.

The fixed fee arrangement not only has the benefit of predictability; it also permits easier cost justification. Customer payback demands are quite stringent. A summary of a typical ARDIS field service cost/benefit analysis is shown in Appendix I. Savings must be defensible from a dozen displaceable costs; the payback certainly has to occur within two years (in the Appendix I example the break-even point is reached in 1.38 years).

With a lease form of payout the first year of the business case does not have to absorb the full capital cost of devices. This distribution of costs effectively broadens the justification time horizon, something vertical customers concentrate on with great intensity.

15.3 HORIZONTAL APPLICATION EXAMPLES: E-MAIL

15.3.1 The Search for the Killer App

History's first electrical mail system was the telegraph. Western Union coined that name in 1851[11] and still has the registered trademark for the term *electronic mail.*

E-mail definitions alone could fill this chapter. In a broad sense it includes facsimile, and both computer-based text and voice messaging. Long-term technology trends will permit a user to communicate regardless of source, format, or destination. Wireless is a long way from achieving these goals, but the drive *has* begun.

This section will focus on what has been, or is just about to be, accomplished: essentially text transmission. It will make one distinction: *messaging* will be the term applied for real-time, person-to-person communication. The term *e-mail* is reserved for systems employing a mailbox. With this capability recipients need not be present when mail is sent to them. All messages are collected in the user's electronic mailbox; users either sign on, or more recently are hunted down, to receive their mail.

Computer-based messaging evolved in the 1970s and began to pick up speed with the development of the (then) low-cost IBM 3270 interactive display system and a plethora of host-based application programs. In the late 1970s, prior to the advent of the IBM PC, dedicated word processing machines began to be affordable (the IBM Displaywriter was just one example).

The PC itself pushed the pace even more intensely. Ironically, within IBM the PC was initially in short supply as production was allocated to external users. In its place the 3270 interactive display based system, tied to a mainframe executing PROFS (Professional Office System), swept through IBM. By 1982 more than 1,000 large mainframes within IBM, and 50 from interested universities, were operational on "VNET" in 29 countries,[14] the largest e-mail system in existence. By 1985 the capability had been extended to IBM employee homes. The AT command set was employed on dial-in modems, with PCs performing the role of 3270 emulator, aided by proprietary data compression algorithms.

E-mail is one of the most pervasive general applications in business with nearly 15 million U.S. users in late 1992.[12] In 1991 the e-mail user count attracted the attention of Ericsson GE, which used its influence to move RAM toward a horizontal marketing strategy: "E-mail will be a vital component of the Mobitex strategy. The vertical markets are certainly important but there's likely to be a faster and potentially greater payoff from concentrating on horizontal applications. The sales cycle for horizontal applications is significantly shorter than for vertical markets."[13]

In May 1992 DEC announced it would develop a version of its client/server Mobilizer software for its 3 million All-in-1 integrated office system customers.[15] The wireless carrier would be RAM. The system would be operational 1Q93. It actually went into internal beta test at Bell South Mobility in June 1993,[16] but mutual confidence (and admiration) remains high.

In November 1992 RAM staged a "coming out party" where several key e-mail alliances were announced.[17] Just one of those alliances produced "Viking Express," an HP/Mobidem/RadioMail-based offering. Euphoria reigned: "they're attacking the beachfront [sic] of Corporate America"[18]; "This is an early warning to CIOs."[19] "Viking Express proves that E-mail is indeed a 'killer application' for wireless data networks."[20]

Possibly, but sometimes unnoticed is the fact that the airtime money is not retail. RadioMail buys airtime blocks from RAM wholesale and rebundles it into 50-word, 100-message, $89 lumps.[21] The original price target was $50 per month.[22] Note, however, that the message traffic is counted both when sending and receiving;[23] it doesn't take long to run the message count up with this form of accounting.

In any case Viking Express has not yet produced a RAM e-mail customer stampede. In November 1992 RadioMail announced[24] that it "had more than 200 paying customers on its e-mail gateway." As of June 21, 1993, RadioMail had "assigned approximately 1,500 mailboxes,"[25] and not all of them were for

RAM (many, possibly most, of RadioMail's customers are on Skylink).

15.3.2 Limiting the Search

From 50,000 feet Electronic Mail Systems has been simplistically visualized[26] as in Figure 15-5.

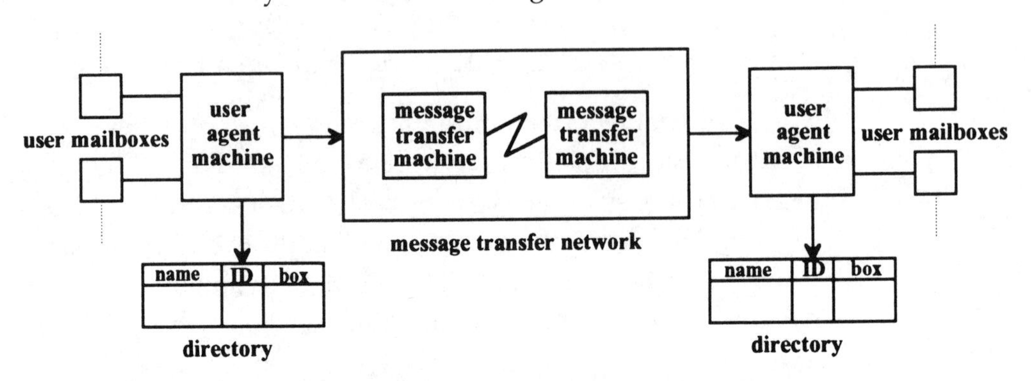

Figure 15-5 *Electronic Mail Overview*

The *user agent* machine provides text editing and presentation services to the end user. It also is responsible for security, priority levels, delivery notification, and distribution subsets.

The *message transfer* machine performs the actual routing. It has responsibility for the store and forward path, channel security, and the actual mail routing. At least eight X.400-X.430 recommendations cover these three functions.

The general representation shown here is symmetrical, but client-server configurations exist as well. DEC's Mobilizer/RF is one such example. The client is a DOS-based PC with a wireless modem; the server is a host VAX that contains the All-in-1 mailboxes.

15.3.3 Add-Ins, Gateways, and Servers

15.3.3.1 Standardization Muddles

If the simple diagram depicted in Figure 15-5 were to be replaced with real-world examples of possible wireless configurations, the variations would cover a wall. Both one-way and two-way systems exist, and many more vendor-specific variants have been announced. There is little standardization; the Lotus VIM versus Microsoft MAPI versus XAPIA Common Messaging Call API "mud fest" interface arguments[27] are a signal of the depth of this problem.

An example of possible application variations is an extract from Ex Machina's suite of offerings that permit selective delivery of messages to paging systems:

Notify!

a. Shipped as a Macintosh System 7.0 System Extension (an "appe"), it resides in the Extensions folder within the System folder. It can be controlled by any application that can send it Apple events or that can utilize external code resources. It takes the data plus the name or other addressing, information in the message and connects (via the Macintosh Communications Tool Box) into a paging system. The Notify! API is documented in the Object Model. . . .
—OR—

b. Consists of a DOS 5.0 (or higher) TSR program. It is user popup and executable. Applications make a simple function call (using the royalty-free glue routines) passing the name. . . .
— OR —

c. Consists of a Windows 3.1 (or higher) DLL server and direct messaging application. Any application capable of sending DDE messages, whether under program control or via user-level scripting. . . .

Got it? Feel comfortable? Is it now clear that e-mail development will lead to "short cycle applications"? No? And this example is not particularly complex as e-mail goes.

15.3.3.2 Gateway Example: RadioMail

Let us focus on one gateway that is operational, and that permits wireless e-mail to be sent to paging systems and to be sent/received from either RAM or ARDIS: RadioMail.

Founded in 1988 as Anterior Technology, a decisive move to wireless e-mail gateway services was underway by 1991. The following year the company's name was changed to RadioMail to reflect its principal thrust. In 1993 it received a critical cash infusion of $3 million as 2M Invest and Motorola purchased a minority stake.[28]

Usually pictured with scores of lines entering its "cloud" from AT&T Easylink, CompuServe, MCI Mail, and others, RadioMail often uses the "network of networks," Internet, for its connections and transport. Similarly, it uses a single feed from Comtex[29] to obtain multiple news services, weather and financial data for its AgentSee/NewsFactory offering, and National Dispatch Center to connect with 220 one-way (paging) services.[30] But the number and types of connections are so rich that they must be grouped in some logical manner as shown in Figure 15-6.

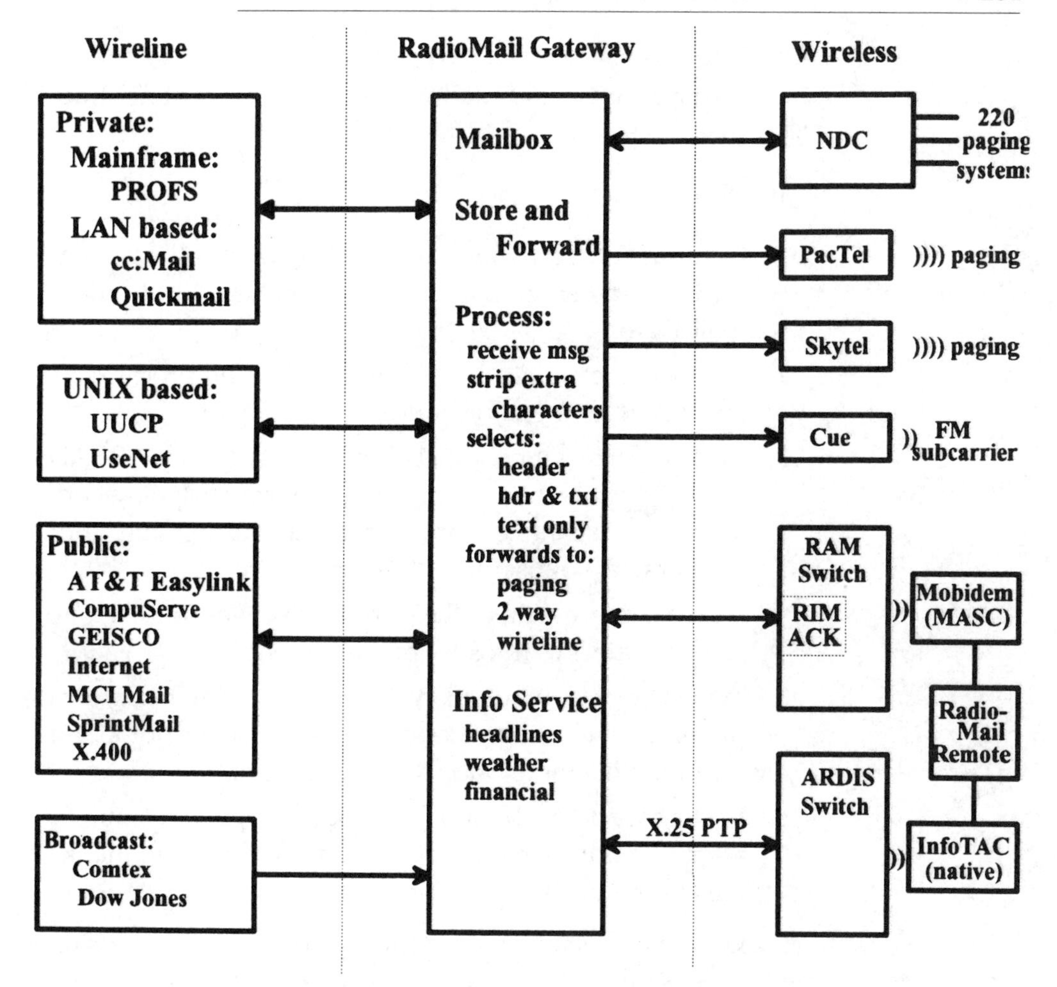

Figure 15-6 *RadioMail Gateway*

Note that both the Mobidem and the InfoTAC are supported in "native"—not AT—mode.

Seemingly trivial tasks such as "strip extra characters" are, in fact, functions critical to the success of a packet switched gateway.

Example: Appendix J is a two-page printout from an actual e-mail inquiry that found mail waiting, a message consisting of 50 text characters.

After the sign-on:

1. CompuServe responded with 773 characters (including much advertising) that reported "You have Electronic mail Waiting."
2. Eight characters (go mail <Enter>) inbound produced a 206-character menu response.
3. Two characters inbound (1<Enter>) selected the "READ MAIL" option; the CompuServe/Internet response consisted of 503 control characters, mostly addressing, and 50 characters of user text: "Hi Jim, did you. . . ."
4. A single depression of the <Enter> key was followed by 273 characters of (mostly) menu options.
5. Two characters inbound (1<Enter>) instructed CompuServe to delete the message; 104 characters outbound informed me there was "No mail waiting."
6. Four characters inbound (off<Enter>) logged me out after 89 more characters outbound, essentially to thank me and report the usage time.

This kind of character traffic is simply not a problem for circuit switched cellular. The entire transaction took less than a minute (probably 20–30 seconds) but hundreds of characters flowed on many continually reversing packets. The packet switched bill for this transaction would be oppressive without the gateway stripping unnecessary traffic. The question: How good a job of this do each of the many gateway alternatives achieve?

The most obvious technique for suppressing unwanted traffic is to prevail on the provider to block it. Most services listen courteously, even enthusiastically, to these requests; some promise to take care of it. None of the services I use ever have.

As a RadioMail user I listen to the advice from the users group[31] ("RMUG"):

1. Buy more memory for the HP.
2. Get one e-mail address:
   a. Use the "autoforward" option available on many systems (MCIMail).
   b. Use rules and routing filters on other systems.
   c. Accept the fact that some systems offer no relief.
3. Fill out the address book with structured aliases.
4. Use cut and paste to move information in and out of messages.
5. Learn and use HP Personal Information Management (PIM) tools if you want to get productivity gains.

Reasonable people may well disagree with RMUG's assessment that "it's not a techie toy." This is tough, often nonintuitive, often clumsy work with many limitations, including message size. The use of autoforwarding, for example, forces "polling" to occur between the Internet and the e-mail system, which can delay messages for hours.[32] Further, RadioMail itself "uses a new, proprietary transport protocol (layer 4)."[33] The RadioMail design has flushed errors in Mobidem microcode, which uses the first three levels of the Mobitex protocol stack to perform functions such as entering power saving mode.

RadioMail is extraordinarily interesting; it is not yet ready for fast rollout horizontal growth.

15.3.3.3 Wireless E-Mail Service Example: RAM

RAM has been engaged in e-mail alliance building since 1991. So much has been published about the alliances, and so much emphasis on interesting events such as the NewsCom demo at the 1992 Democratic National Convention,[34] that it is often hard to focus on what actually exists. The operational linkages (solid lines) plus several planned for 1993 (dotted) are shown in Figure 15-7.

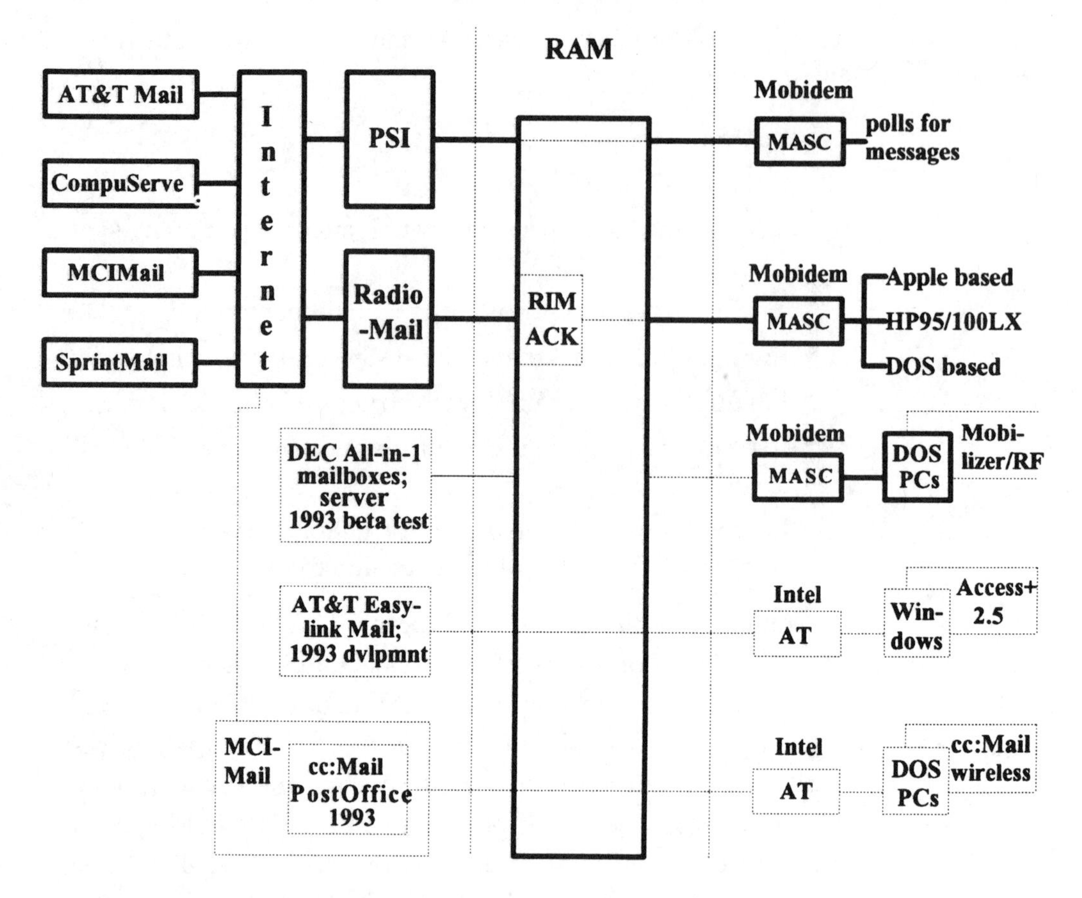

Figure 15-7 *RAM E-Mail Alliances*

This simplified diagram does not include several alliances that lack date information:

1. TEKnique's UNIX-based TransRmail e-mail servers for Canada and the United Kingdom
2. Telepartner International's comarketing arrangement for Mobi/3270 access to IBM-compatible hosts
3. Simware's plan for SimVision/PC access to PROFS/OfficeVision IBM hosts
4. WordPerfect 4.0 gateway plans

But it is quite clear that RAM is doing an outstanding job of attacking the complex problem of electronic communication.

Technical surprises abound and RAM has been batting them down one by one. Example: The Mobitex network does not provide a positive acknowledgment as found in ARDIS. The dispatching user is informed that a remote subscriber did *not* receive a message—perhaps hours later.[35] Thus, RAM had to engage Research-in-Motion to place its Mobilib Plus API at the switch so RadioMail could know the status of the traffic. There have been frustrating Mobidem microcode problems (as there have been with the InfoTAC for ARDIS).

E-mail Net:

1. Existing services are expensive., RadioMail's message lengths are short (50 characters), perhaps reflecting its original customer base: paging. The $89 fee is paid for each message sent or received. This quickly becomes hard to cost justify:[36] "GRiD tested RadioMail for executives but couldn't justify the expense." Further, "employees don't want to learn how to use [RadioMail]." Instead, GRiD uses a toll-free telephone for employees to access Lotus cc:Mail for $30/month/employee.

   Claims that "RadioMail is fun."[37] are simply not defensible in a business case.
2. Existing and planned services require a high level of technical knowledge.
3. Current implementations are clumsy.
4. The fast rollout envisioned in 1991 has not materialized.

This is a tough, slow business.

REFERENCES

[1] Janet Coles, PacTel Information Service Center, 6-22-93.
[2] Carl Arons, RAM Chief Executive Officer, Mobile '93 presentation.
[3] *RAM Hard Data*, Fall 1992.
[4] ARDIS Lexington Conference, 7-27-93.
[5] *Mobile Phone News*, 4-26-93.
[6] ARDIS Lexington Conference, 7-27-93.

[7] Representative statistics extracted from Yankee Watch, Wireless/Mobile Communications, 2Q93. Reprints distributed by ARDIS at Lexington Conference, 7-27-93.

[8] ARDIS Lexington Conference, 7-27-93.

[9] *Big Blues: The Unmaking of IBM*, p. 128.

[10] *SMRS Market Planning Analysis: New York City*, Table 6, p. 10.

[11] From *Telecommunications for Management*, Charles T. Meadow and Albert S. Tedesco; Ruann Pengov (GEISCO) reference.

[12] Electronic Mail Association report, *Information Week*, 11-2-92.

[13] William Frezza, Ericsson GE Director of Strategic Sales, *Mobile Data Report*, 12-2-92.

[14] Lewis M. Branscomb, IBM Chief Scientist, Keynote Address, IEEE INFOCOM '83.

[15] *Telecommunications Alert*, 5-22-92.

[16] *Edge On & About AT&T*, 6-21-93.

[17] *Electronic Messaging News*, 11-11-92.

[18] Jay Baylock, analyst, Gartner Group, *Information Week*, 11-2-92.

[19] Ibid., quoting William Frezza, Ericsson GE Mobile Communications.

[20] *Mobile Data Report*, 12-7-92.

[21] Ibid.

[22] *Mobile Data Report*, 10-21-91.

[23] *Mobile Data Report*, 2-1-93.

[24] Geoffrey Goodfellow, RadioMail Chairman, *Mobile Data Report*, 1-9-92.

[25] Geoffrey Goodfellow, *Communications Week*, 6-21-93.

[26] *Computer Networks: Protocols, Standards and Interfaces*, Uyless Black, 1987.

[27] *Electronic Messaging News*, 6-23-93.

[28] *Mobile Phone News*, 4-26-93.

[29] *Mobile Data Report*, 10-21-91.

[30] Ibid.

[31] Jim Opfer, 2311 Berendo Avenue, Torrance, CA, 90502; or try: opfer @ radiomail.

[32] *Mobile Data Report*, 3-1-93.

[33] Ibid.

[34] *Data Channels*, 8-3-92.

[35] *Mobile Data Report*, 12-26-91.

[36] Jerry Keeran, AST Manager of Communication Products, *Mobile Data Report*, 8-2-93.

[37] Geoffrey S. Goodfellow, RadioMail Chairman, *Edge On & About AT&T*, 11-2-92.

CHAPTER

16

SERVICE PROVIDER CONCERNS

16.1 APPLICATION SELECTION INTRODUCTION

Chapter 11 demonstrated one technique for service providers to calculate supportable messages per second given that they are able to make reasonable assumptions concerning the:

1. Average C/N in the coverage area
2. Nominal speed (in mph) of the "typical" target
3. Average length of the message in octets

This technique will be expanded, using CDPD V0.8 as an example, to demonstrate the difficulty the service provider will encounter in selecting the "right" application for the network.

16.2 CDPD MESSAGE-PER-SECOND TABLES

16.2.1 Hopping Impact

Prior examples focused exclusively on dedicated base stations, a legitimate configuration alternative for CDPD. However, the system was designed with spectrum sharing as a principal goal: Data can be inserted on the airwaves when voice is idle.

Unfortunately, highly granular (say, one-second interval) statistics on voice cellular operation are rare. Carriers know their busiest cells and are quite capable of reporting whether or not a cell exceeded a performance threshold in some busy hour. Most do not know whether a brief (say, 30 seconds) period of intense voice activity occurred in an otherwise uneventful hour. This is not particularly critical to voice; it can be a major problem for data.

As a result, JFD Associates wrote a Pascal model to create a test *voice* scenario for data-channel hopping. The model explanation (and limitations) can be found in Appendix K. A voice cell was assumed to have 19 channels, assigned round robin, working close to, but not exceeding, 2 percent Erlang-B blocking limits on an hourly basis. Hopping was preemptive; if a better channel position became free, hopping occurred.

A profile was established in which the average voice call length was about 2 minutes, including setup/tear-down. The "exogenous shock" situation was one peak period of 60 seconds duration in an hour. During this intense peak, voice calls were attempted at the rate of about one per second. Sample model results (rounded) are shown in Table 16-1.

Table 16-1 *Voice Simulation Model Results: Hopping Input*

1.	Voice calls per hour	341	11.5 Erlangs or
2.	Average voice call duration (seconds)	121	1.2% blocking
3.	Planned hops/hour	169	
4.	Too-busy-for-data busy seconds	90	
5.	Unique data-blocked periods	13	

It is important to note that the data-busy seconds occur as a result of voice bursts. During these intervals no packets at all can be transmitted. If the application is, say, credit card verification the user will simply have to wait. It is key to understand that this is a *voice*, not data, driven phenomenon. There can be a single data user on the system and that user will be blocked if voice traffic peaks as described.

All devices are assumed to be working in a routing area subdomain: All base stations are under the control of a single MD-IS; there is no redirection and forwarding.

There was no allowance for hops forced by, say, cochannel interference thresholds; virtually all hops were planned. Further, each planned hop was assumed to consume only one second of channel time for the necessary drying up of the forward-channel message stream, selection of the new channel, examination of its signal strength, achieving synchronization, evaluation of the block error rate, issuance of request, open, activate, confirm, procedures, and so on.

The 13 data-blocked events were assumed to consume two seconds of airtime per event to account for abort/recover.

The total lost capacity is as shown in Table 16-2.

Table 16-2 *Lost Data Capacity: Hopping*

1.	Planned hops	(169 × 1)	169.0
2.	Unplanned aborts	(13.3 × 2)	26.6
3.	Data-busy time		90.3
			285.9 ÷ 3600 = ~8%

16.2.2 Outbound Hopping Performance

The impact of hopping on the forward channel was calculated by taking 8 percent of the channel away and assigning it to hop overhead tasks before any capacity calculations began. The error-free potential was reduced by the impact of errors as explained in Chapter 11. The summary results for a 70 percent utilized channel are shown in Table 16-3 in terms of both messages/second and sample active-user capacities.

Table 16-3 *CDPD V0.8 Active User Potential: Outbound Hopping Channel (with errors)*

Octets	Error-free msg/sec	w/ Hop o'head	Useful percent	w/ Error msg/sec	User rate			
					1/60′	1/45′	1/30′	1/15′
5	16.00	14.72	0.947	13.94	50168	37626	25084	12542
10	16.00	14.72	0.947	13.94	50168	37626	25084	12542
15	16.00	14.72	0.947	13.94	50168	37626	25084	12542
20	16.00	14.72	0.947	13.94	50168	37626	25084	12542
25	10.67	9.81	0.896	8.80	31667	23750	15834	7917
30	10.67	9.81	0.896	8.80	31667	23750	15834	7917
35	10.67	9.81	0.896	8.80	31667	23750	15834	7917
40	10.67	9.81	0.896	8.80	31667	23750	15834	7917
45	10.67	9.81	0.896	8.80	31667	23750	15834	7917
50	10.67	9.81	0.896	8.80	31667	23750	15834	7917
55	10.67	9.81	0.896	8.80	31667	23750	15834	7917

Table 16-3 *CDPD V0.8 Active User Potential (continued)*

Octets	Error-free msg/sec	w/ Hop o'head	Useful percent	w/ Error msg/sec	User rate			
					1/60′	1/45′	1/30′	1/15′
60	8.00	7.36	0.849	6.25	22492	16869	11246	5623
65	8.00	7.36	0.849	6.25	22492	16869	11246	5623
70	8.00	7.36	0.849	6.25	22492	16869	11246	5623
75	8.00	7.36	0.849	6.25	22492	16869	11246	5623
80	8.00	7.36	0.849	6.25	22492	16869	11246	5623
85	8.00	7.36	0.849	6.25	22492	16869	11246	5623
90	8.00	7.36	0.849	6.25	22492	16869	11246	5623
95	6.40	5.89	0.804	4.74	17052	12789	8526	4263
100	6.40	5.89	0.804	4.74	17052	12789	8526	4263
105	6.40	5.89	0.804	4.74	17052	12789	8526	4263
110	6.40	5.89	0.804	4.74	17052	12789	8526	4263
115	6.40	5.89	0.804	4.74	17052	12789	8526	4263
120	5.33	4.91	0.840	4.12	14832	11124	7416	3708
125	5.33	4.91	0.840	4.12	14832	11124	7416	3708
130	5.33	4.91	0.840	4.12	14832	11124	7416	3708
135	5.33	4.91	0.840	4.12	14832	11124	7416	3708
140	5.33	4.91	0.840	4.12	14832	11124	7416	3708
145	5.33	4.91	0.840	4.12	14832	11124	7416	3708
150	4.57	4.21	0.851	3.58	12883	9663	6442	3221
155	4.57	4.21	0.851	3.58	12883	9663	6442	3221
160	4.57	4.21	0.851	3.58	12883	9663	6442	3221
165	4.57	4.21	0.851	3.58	12883	9663	6442	3221
170	4.57	4.21	0.851	3.58	12883	9663	6442	3221
175	4.57	4.21	0.851	3.58	12883	9663	6442	3221

Table 16-3 *CDPD V0.8 Active User Potential (continued)*

Octets	Error-free msg/sec	w/ Hop o'head	Useful percent	w/ Error msg/sec	User rate			
					1/60′	1/45′	1/30′	1/15′
180	4.57	4.21	0.851	3.58	12883	9663	6442	3221
185	4.00	3.68	0.848	3.12	11232	8424	5616	2808
190	4.00	3.68	0.848	3.12	11232	8424	5616	2808
195	4.00	3.68	0.848	3.12	11232	8424	5616	2808
200	4.00	3.68	0.848	3.12	11232	8424	5616	2808
205	4.00	3.68	0.848	3.12	11232	8424	5616	2808
210	4.00	3.68	0.848	3.12	11232	8424	5616	2808
215	4.00	3.68	0.848	3.12	11232	8424	5616	2808
220	3.56	3.27	0.836	2.73	9841	7381	4921	2460
225	3.56	3.27	0.836	2.73	9841	7381	4921	2460
230	3.56	3.27	0.836	2.73	9841	7381	4921	2460
235	3.56	3.27	0.836	2.73	9841	7381	4921	2460
240	3.56	3.27	0.836	2.73	9841	7381	4921	2460
245	3.20	2.94	0.727	2.14	7705	5779	3852	1926
250	3.20	2.94	0.727	2.14	7705	5779	3852	1926
255	3.20	2.94	0.727	2.14	7705	5779	3852	1926
260	3.20	2.94	0.727	2.14	7705	5779	3852	1926
265	3.20	2.94	0.727	2.14	7705	5779	3852	1926
270	3.20	2.94	0.727	2.14	7705	5779	3852	1926
275	2.91	2.68	0.748	2.00	7210	5408	3605	1803
280	2.91	2.68	0.748	2.00	7210	5408	3605	1803
285	2.91	2.68	0.748	2.00	7210	5408	3605	1803
290	2.91	2.68	0.748	2.00	7210	5408	3605	1803
295	2.91	2.68	0.748	2.00	7210	5408	3605	1803

Table 16-3 *CDPD V0.8 Active User Potential (continued)*

Octets	Error-free msg/sec	w/ Hop o'head	Useful percent	w/ Error msg/sec	User rate			
					1/60′	1/45′	1/30′	1/15′
300	2.91	2.68	0.748	2.00	7210	5408	3605	1803
305	2.91	2.68	0.748	2.00	7210	5408	3605	1803
310	2.67	2.45	0.761	1.87	6723	5042	3362	1681
315	2.67	2.45	0.761	1.87	6723	5042	3362	1681
320	2.67	2.45	0.761	1.87	6723	5042	3362	1681
325	2.67	2.45	0.761	1.87	6723	5042	3362	1681
330	2.67	2.45	0.761	1.87	6723	5042	3362	1681
335	2.67	2.45	0.761	1.87	6723	5042	3362	1681
340	2.67	2.45	0.761	1.87	6723	5042	3362	1681
345	2.46	2.26	0.767	1.74	6256	4692	3128	1564
350	2.46	2.26	0.767	1.74	6256	4692	3128	1564
355	2.46	2.26	0.767	1.74	6256	4692	3128	1564
360	2.46	2.26	0.767	1.74	6256	4692	3128	1564
365	2.46	2.26	0.767	1.74	6256	4692	3128	1564
370	2.29	2.10	0.654	1.37	4947	3710	2474	1237
375	2.29	2.10	0.654	1.37	4947	3710	2474	1237
380	2.29	2.10	0.654	1.37	4947	3710	2474	1237
385	2.29	2.10	0.654	1.37	4947	3710	2474	1237
390	2.29	2.10	0.654	1.37	4947	3710	2474	1237
395	2.29	2.10	0.654	1.37	4947	3710	2474	1237
400	2.29	2.10	0.691	1.45	5229	3922	2614	1307
405	2.13	1.96	0.675	1.32	4767	3575	2384	1192
410	2.13	1.96	0.675	1.32	4767	3575	2384	1192
415	2.13	1.96	0.675	1.32	4767	3575	2384	1192

Table 16-3 *CDPD V0.8 Active User Potential (continued)*

Octets	Error-free msg/sec	w/ Hop o'head	Useful percent	w/ Error msg/sec	User rate			
					1/60′	1/45′	1/30′	1/15′
420	2.13	1.96	0.675	1.32	4767	3575	2384	1192
425	2.13	1.96	0.675	1.32	4767	3575	2384	1192
430	2.13	1.96	0.675	1.32	4767	3575	2384	1192
435	2.00	1.84	0.691	1.27	4578	3433	2289	1144
440	2.00	1.84	0.691	1.27	4578	3433	2289	1144
445	2.00	1.84	0.691	1.27	4578	3433	2289	1144
450	2.00	1.84	0.691	1.27	4578	3433	2289	1144
455	2.00	1.84	0.691	1.27	4578	3433	2289	1144
460	2.00	1.84	0.691	1.27	4578	3433	2289	1144
465	2.00	1.84	0.691	1.27	4578	3433	2289	1144
470	1.88	1.73	0.703	1.22	4383	3287	2192	1096
475	1.88	1.73	0.703	1.22	4383	3287	2192	1096
480	1.88	1.73	0.703	1.22	4383	3287	2192	1096
485	1.88	1.73	0.703	1.22	4383	3287	2192	1096
490	1.88	1.73	0.703	1.22	4383	3287	2192	1096
495	1.78	1.64	0.604	0.99	3554	2666	1777	889
500	1.78	1.64	0.604	0.99	3554	2666	1777	889
505	1.78	1.64	0.604	0.99	3554	2666	1777	889
510	1.78	1.64	0.604	0.99	3554	2666	1777	889

The active-user potential is rate driven. If the average outbound message is, say, 120 octets, the forward channel can support a message rate of 4.12 messages per second with both hopping and error retries. If the average mobile user receives a message once per hour the supportable users are 3600 seconds × 4.12

messages per second = 14,832 users. But if the average message rate is once every 15 minutes supportable users drop to 900 × 4.12 = 3,708.

16.2.3 Inbound Hopping Performance

An intermediate calculation was required on the inbound channel. The hopping overhead of 8 percent was first removed from total time available. The *G* was recomputed at 70 percent on the 92 percent remaining capacity. Then *S*, the post-contention throughput, was calculated before errors were encountered.

The reduced message-per-second potential was then computed with the impact of error retries. The summary results, which also include the active user potentials, are tabulated in Table 16-4. Note that the inbound channel at a message length of 120 octets has roughly one-half the traffic potential as the outbound channel at that same length.

Table 16-4 *CDPD V0.8 Active User Potential: Inbound Hopping Channel (with errors)*

Octets	Error-free msg/sec	w/ Hop o'head	Useful percent	w/ Error msg/sec	User rate			
					1/60′	1/45′	1/30′	1/15′
5	6.88	6.57	0.910	5.98	21535	16152	10768	5384
10	6.88	6.57	0.910	5.98	21535	16152	10768	5384
15	6.88	6.57	0.910	5.98	21535	16152	10768	5384
20	6.88	6.57	0.910	5.98	21535	16152	10768	5384
25	5.24	5.00	0.844	4.22	15189	11392	7595	3797
30	5.24	5.00	0.844	4.22	15189	11392	7595	3797
35	5.24	5.00	0.844	4.22	15189	11392	7595	3797
40	5.24	5.00	0.844	4.22	15189	11392	7595	3797
45	5.24	5.00	0.844	4.22	15189	11392	7595	3797

Table 16-4 *CDPD V0.8 Active User Potential (continued)*

Octets	Error-free msg/sec	w/ Hop o'head	Useful percent	w/ Error msg/sec	User rate			
					1/60′	1/45′	1/30′	1/15′
50	5.24	5.00	0.844	4.22	15189	11392	7595	3797
55	5.24	5.00	0.844	4.22	15189	11392	7595	3797
60	4.10	3.91	0.784	3.06	11022	8267	5511	2756
65	4.10	3.91	0.784	3.06	11022	8267	5511	2756
70	4.10	3.91	0.784	3.06	11022	8267	5511	2756
75	4.10	3.91	0.784	3.06	11022	8267	5511	2756
80	4.10	3.91	0.784	3.06	11022	8267	5511	2756
85	4.10	3.91	0.784	3.06	11022	8267	5511	2756
90	4.10	3.91	0.784	3.06	11022	8267	5511	2756
95	3.45	3.29	0.729	2.40	8633	6475	4316	2158
100	3.45	3.29	0.729	2.40	8633	6475	4316	2158
105	3.45	3.29	0.729	2.40	8633	6475	4316	2158
110	3.45	3.29	0.729	2.40	8633	6475	4316	2158
115	3.45	3.29	0.729	2.40	8633	6475	4316	2158
120	2.92	2.78	0.781	2.17	7814	5861	3907	1954
125	2.92	2.78	0.781	2.17	7814	5861	3907	1954
130	2.92	2.78	0.781	2.17	7814	5861	3907	1954
135	2.92	2.78	0.781	2.17	7814	5861	3907	1954
140	2.92	2.78	0.781	2.17	7814	5861	3907	1954
145	2.92	2.78	0.781	2.17	7814	5861	3907	1954
150	2.92	2.78	0.801	2.23	8011	6008	4005	2003
155	2.57	2.45	0.801	1.96	7066	5300	3533	1767
160	2.57	2.45	0.801	1.96	7066	5300	3533	1767
165	2.57	2.45	0.801	1.96	7066	5300	3533	1767

Table 16-4 *CDPD V0.8 Active User Potential (continued)*

Octets	Error-free msg/sec	w/ Hop o'head	Useful percent	w/ Error msg/sec	User rate			
					1/60′	1/45′	1/30′	1/15′
170	2.57	2.45	0.801	1.96	7066	5300	3533	1767
175	2.57	2.45	0.801	1.96	7066	5300	3533	1767
180	2.57	2.45	0.801	1.96	7066	5300	3533	1767
185	2.57	2.45	0.800	1.96	7063	5297	3531	1766
190	2.26	2.16	0.800	1.73	6213	4659	3106	1553
195	2.26	2.16	0.800	1.73	6213	4659	3106	1553
200	2.26	2.16	0.800	1.73	6213	4659	3106	1553
205	2.26	2.16	0.800	1.73	6213	4659	3106	1553
210	2.26	2.16	0.800	1.73	6213	4659	3106	1553
215	2.26	2.16	0.800	1.73	6213	4659	3106	1553
220	2.26	2.16	0.788	1.70	6113	4585	3057	1528
225	2.05	1.95	0.788	1.54	5540	4155	2770	1385
230	2.05	1.95	0.788	1.54	5540	4155	2770	1385
235	2.05	1.95	0.788	1.54	5540	4155	2770	1385
240	2.05	1.95	0.788	1.54	5540	4155	2770	1385
245	2.05	1.95	0.655	1.28	4605	3453	2302	1151
250	1.85	1.76	0.655	1.15	4151	3113	2075	1038
255	1.85	1.76	0.655	1.15	4151	3113	2075	1038
260	1.85	1.76	0.655	1.15	4151	3113	2075	1038
265	1.85	1.76	0.655	1.15	4151	3113	2075	1038
270	1.85	1.76	0.655	1.15	4151	3113	2075	1038
275	1.85	1.76	0.684	1.20	4334	3251	2167	1084
280	1.85	1.76	0.684	1.20	4334	3251	2167	1084
285	1.68	1.60	0.684	1.10	3946	2959	1973	986

Table 16-4 *CDPD V0.8 Active User Potential (continued)*

Octets	Error-free msg/sec	w/ Hop o'head	Useful percent	w/ Error msg/sec	User rate			
					1/60′	1/45′	1/30′	1/15′
290	1.68	1.60	0.684	1.10	3946	2959	1973	986
295	1.68	1.60	0.684	1.10	3946	2959	1973	986
300	1.68	1.60	0.684	1.10	3946	2959	1973	986
305	1.68	1.60	0.684	1.10	3946	2959	1973	986
310	1.68	1.60	0.702	1.13	4054	3040	2027	1013
315	1.68	1.60	0.702	1.13	4054	3040	2027	1013
320	1.56	1.49	0.702	1.05	3764	2823	1882	941
325	1.56	1.49	0.702	1.05	3764	2823	1882	941
330	1.56	1.49	0.702	1.05	3764	2823	1882	941
335	1.56	1.49	0.702	1.05	3764	2823	1882	941
340	1.56	1.49	0.702	1.05	3764	2823	1882	941
345	1.56	1.49	0.713	1.06	3819	2864	1910	955
350	1.56	1.49	0.713	1.06	3819	2864	1910	955
355	1.44	1.37	0.713	0.98	3525	2644	1763	881
360	1.44	1.37	0.713	0.98	3525	2644	1763	881
365	1.44	1.37	0.713	0.98	3525	2644	1763	881
370	1.44	1.37	0.581	0.80	2876	2157	1438	719
375	1.44	1.37	0.581	0.80	2876	2157	1438	719
380	1.35	1.29	0.581	0.75	2698	2024	1349	675
385	1.35	1.29	0.581	0.75	2698	2024	1349	675
390	1.35	1.29	0.581	0.75	2698	2024	1349	675
395	1.35	1.29	0.581	0.75	2698	2024	1349	675
400	1.35	1.29	0.624	0.80	2895	2172	1448	724

Table 16-4 *CDPD V0.8 Active User Potential (continued)*

Octets	Error-free msg/sec	w/ Hop o'head	Useful percent	w/ Error msg/sec	User rate			
					1/60′	1/45′	1/30′	1/15′
405	1.35	1.29	0.608	0.78	2822	2116	1411	705
410	1.35	1.29	0.608	0.78	2822	2116	1411	705
415	1.26	1.20	0.608	0.73	2632	1974	1316	658
420	1.26	1.20	0.608	0.73	2632	1974	1316	658
425	1.26	1.20	0.608	0.73	2632	1974	1316	658
430	1.26	1.20	0.608	0.73	2632	1974	1316	658
435	1.26	1.20	0.629	0.76	2724	2043	1362	681
440	1.26	1.20	0.629	0.76	2724	2043	1362	681
445	1.26	1.20	0.629	0.76	2724	2043	1362	681
450	1.19	1.14	0.629	0.72	2575	1932	1288	644
455	1.19	1.14	0.629	0.72	2575	1932	1288	644
460	1.19	1.14	0.629	0.72	2575	1932	1288	644
465	1.19	1.14	0.629	0.72	2575	1932	1288	644
470	1.19	1.14	0.646	0.73	2642	1981	1321	660
475	1.12	1.07	0.646	0.69	2484	1863	1242	621
480	1.12	1.07	0.646	0.69	2484	1863	1242	621
485	1.12	1.07	0.646	0.69	2484	1863	1242	621
490	1.12	1.07	0.646	0.69	2484	1863	1242	621
495	1.12	1.07	0.536	0.57	2062	1546	1031	515
500	1.12	1.07	0.536	0.57	2062	1546	1031	515
505	1.12	1.07	0.536	0.57	2062	1546	1031	515
510	1.07	1.02	0.536	0.54	1961	1471	981	490

16.2.4 Performance Tables Demonstrated

It is clear from the prior tables that CDPD would be inbound limited if all messages were of equal size. Fortunately, large classes of applications, particularly those with an inquiry/response profile, have shorter inbound message lengths. A generalized technique for using the tables to obtain the limits is:

1. Determine the average interarrival rate for both inbound and outbound messages in seconds. For example, if the inbound rate is 3 per hour, the message rate is one every 1200 seconds.
2. Use the average message length to look up the message-per-second rate in Tables 16-3 and 16-4. For example, if the outbound message length averages 120 octets the Table 16-3 usable rate is 4.12 messages per second; if the inbound message length averages 75 octets the Table 16-4 usable rate is 3.06 messages per second.
3. Multiply the interval in seconds times the message-per-second rate to obtain the supportable number of active users. In our examples:

   Outbound: $1200 \times 4.12 = 4{,}944$
   Inbound: $1200 \times 3.06 = 3{,}672$

16.3 STAND-ALONE APPLICATION EXAMPLES

Although no two applications are identical, certain classes of applications have similar traffic profiles. Five representative profiles are shown in Table 16-5. They range from unsuitable *for the service provider* (IBM CICS/IMS), to severely output limited (Motorola DBI), to sharply input limited (Taxi Dispatch). None have balanced output versus input—which is quite normal—thus capacity is often wasted. The best balanced application is field service, which has been well penetrated by ARDIS.

Table 16-5 *Representative Application Profiles*

	Step 1		Step 2		Step 3	Active users
	Message per hour	**Seconds interval**	**Message length**	**Message per sec**		
Field Service:						
** outbound:	5.0	720	100	4.74	4.74 × 720 =	3413
inbound:	3.5	1029	45	4.22	4.22 × 1029 =	4342
Taxi Dispatch:						
** outbound:	24.0	150	15	13.94	13.94 × 150 =	2091
inbound:	36.0	100	15	5.98	5.98 × 100 =	598
Public Safety:						
** outbound:	33.0	109	100	4.74	4.74 × 109 =	517
inbound:	14.0	257	17	5.98	5.98 × 257 =	1537
Motorola DBI:						
** outbound:	4.0	900	2000	.25	.25 × 900 =	225
inbound:	4.0	900	20	5.98	5.98 × 900 =	5382
IBM CICS/IMS:						
** outbound:	60.0	60	750	.66	.66 × 60 =	40
inbound:	60.0	60	40	4.22	4.22 x 60 =	253
** limiting direction						

To demonstrate the problem that a profile such as IBM CICS/IMS brings to the service provider (not the user) assume that packet switched airtime prices have been beaten down to $.005/100 octets as described in Chapter 11.

Then the user's costs will be:

$$\begin{array}{rl} 750 \div 100 = 8 \text{ priced segments} \times \$.005 = & \$.040 \text{ (outbound)} \\ 40 \div 100 = 1 \text{ priced segment} \times \$.005 = & \underline{\ \ .005} \text{ (inbound)} \\ & \$.045 \times 60 \text{ two-way transactions} = \$2.70 \text{ hour} \end{array}$$

What a neat deal! Low packet prices, or flat-rate pricing, is sure to attract the high-volume user.

But because of outbound channel limitations only 40 such users are sustainable at any one time. The outbound channel, derated for hopping—and prudently held to 70 percent utilization—will be sustaining:

$$\frac{60 \text{ transactions/user hour} \times 40 \text{ users} \times 750 \text{ user octets} \times 8 \text{ bit}}{3600 \text{ seconds per hour}} = 4000 \text{ bps}$$

which is reasonable.

The attraction for the user is a problem for the service provider. The $2.70 per user × 40 users becomes a total hourly revenue of $108. This is not such a neat deal. Thus, the prices must be higher, or cocked to penalize the long message, or this class of user must be refused access to the system (that's hard).

Taxi dispatch is a less pathological case, and doubly interesting because the technology has been widely accepted in private systems. Many of these systems will be nearing end-of-life when CDPD is broadly available. At the target price it can be extremely attractive to the user: $.30 per hour. However, with an inbound limit of 598 users the best peak-hour revenue potential is much better than CICS/IMS for the service provider but considerably less than voice cellular:

1. *Voice:* with Erlang-B, 2% blocking, and 19 channels, average channel utilization is ~60%. Assuming $.38/minute voice revenue, as in Chapter 3, a peak hour returns:

19 channels × 60 minutes × 60% utilization × $.38

= $259.92

2. *Data:* assuming $.005 per 100 octets and taxi dispatch receiving absolutely no granularity break on its 15-octet messages:

Outbound: 598 users × 24 msgs/hour × $.005 =	$ 71.76
Inbound: 598 users × 36 msgs/hour × $.005 =	107.64
	$179.40

16.4 BALANCING THE APPLICATION MIX

Assume that public safety users, who are outbound limited, can be overlayed on taxi dispatch. Wasted capacity in both applications would be filled with revenue-producing traffic.

Consider that both applications have single CDPD block inbound messages in a channel capable of handling 5.98 messages per second at this length. The average taxi dispatch user generates one message every 100 seconds, while the public safety counterpart generates a message every 257 seconds. Thus for every unallocated (or deliberately unsold) taxi dispatch user, 2.57 public safety users can be added. This approach permits a net add of *total* users because public safety is less message intensive. A snapshot in 50-unit steps as taxi dispatch users are curtailed is shown in Table 16-6.

Table 16-6 *Taxi Dispatch versus Public Safety Active User Tradeoff*

T.D.	P.S.	Total
598	0	598
548	129	677
498	257	755
448	386	834
398	514	912
348	643	991
298	771	1069
248	900	1148
198	1028	1226
148	1157	1305
98	1285	1383
48	1414	1462
0	1537	1537

The outbound channel tradeoff can be calculated in a similar manner. The only complication is the difference in message lengths, which must be solved.

As for inbound, the first step is to compare rates: one message every 150 seconds for taxi dispatch; one message every 109 seconds for public safety. If messages sizes were identical, each taxi dispatch unit sacrificed would permit 109/150 = .727 public safety users to be added.

But the public safety message is longer; the maximum message rate is 4.74 per second versus 13.94 for taxi dispatch. Thus public safety must be further derated on the outbound side:

$$\frac{4.74 \text{ PS msg/sec}}{13.94 \text{ TD msg/sec}} \times \frac{109 \text{ PS interval in sec}}{150 \text{ TD interval in sec}}$$

$$= .247 \text{ PS adds for each TD removal}$$

But these outbound channel taxi dispatch "removals" are phantoms; the application has been constrained by the inbound side; the outbound channel has idle time. Thus, adding public safety causes the overall outbound utilization to rise.

A graphic view is shown in Figure 16-1.

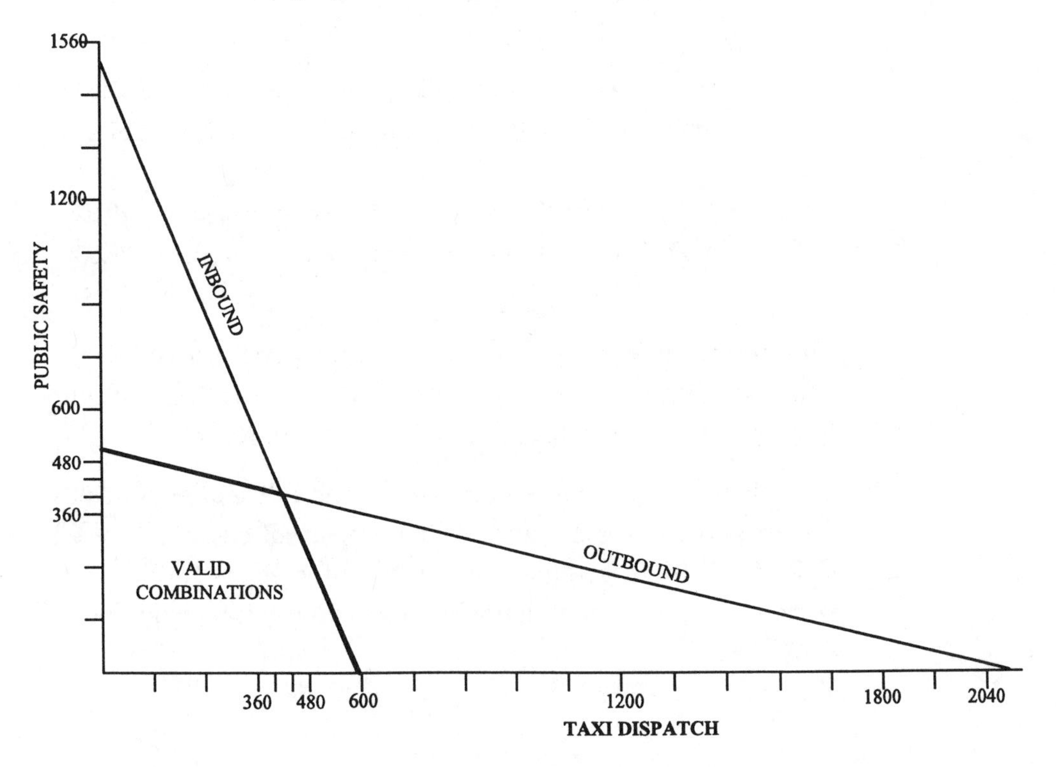

Figure 16-1 *Voice Simulation Model Results: Hopping Input*

This strange graph plots two lines:

1. The unattainable and wasted inbound capacity of public safety (1,537 users) to the inbound-limited taxi dispatch (598 users).

2. The outbound-limited public safety (517 users) to the unattainable, wasted outbound capacity of taxi dispatch.

The area below the intersection represents the mix of active users that will provide a better channel utilization than either application treated separately.

The "sweet spot," the highest revenue yield for the two applications, is 439 taxi dispatch users (not the 598 achievable if standalone) and 409 public safety users (not the 517 theoretically achievable). The results:

Outbound:

Taxi Dispatch: 439 users × 24 msgs/hour × $.005 = $ 52.68
Public Safety: 409 users × 33 msgs/hour × $.005 = 67.49

Inbound:

Taxi Dispatch: 439 users × 36 msgs/hour × $.005 = 79.02
Public Safety: 409 users × 14 msgs/hour × $.005 = 28.63

$ 227.82

This permits packet switched data to approach voice cellular in peak hourly revenue—and further indicates that a price outlook of ~$.005/100 octets, over time, is "about right."

But this mix balancing requires uncommon discipline for sales forces who, by training and inclination, are often happy to sell *anybody*. If the discipline cannot be enforced as traffic rises, packet switched data will not be a particularly interesting business.

APPENDIX

ACRONYM GUIDE

ACK Acknowledgment.

ALOHA The University of Hawaii data radio system featuring unconstrained contention access on the inbound side.

ARDIS Advanced Radio Data Information System: the 50/50 joint venture between IBM and Motorola, founded in 1990 to provide a public airtime service. It employs Motorola protocols and infrastructure.

ARQ Automatic Repeat Request: retransmission of message segments when an error has been detected.

AVL Automatic Vehicle Location.

bps Bits per second.

C/I Carrier-to-Interference ratio in decibels.

CDLC Cellular Data Link Control: a protocol developed by Racal and used by Millicom (among others) to transmit data via cellular connections.

CDPD Cellular Digital Packet Data: the name employed for the system that is able to share cellular spectrum with voice. It can also operate on dedicated spectrum. When sharing, voice has priority. The system is not digital in the sense of being modemless. Its over-the-air modems are GMSK modulated.

CRC Cyclic Redundancy Check: a mathematical operation applied to data about to be transmitted. The result, usually two (CRC-16) or four (CRC-32) octets long is appended. Upon message receipt the same mathematical operation is performed and checked against the CRC. If there is a mismatch, an error has occurred.

CSMA Carrier Sense Multiple Access: the technique used to reduce contention by listening for "busy" indicators before transmitting; an improved throughput technique over ALOHA variations.

DCS Digital Communication System: the IBM name for its field service system. A misnomer (strictly speaking the system was analog), it was folded into ARDIS in 1990.

DRN Data Radio Network: Motorola's first public airtime service, which began in 1986 and was folded into ARDIS in 1990.

ESMR Enhanced SMR: the most noteworth pioneer is Nextel employing MIRS technology.

ETC Enhanced Transmission Cellular: the protocol employed by AT&T Paradyne for cellular data transfer. It claims fast, robust connection at high (14.4 Kbps) transmission speeds.

fax Facsimile machine

FEC Forward Error Correction: usually applied to a class of codes that is able to mathematically repair some damage that may occur during radio transmission.

G The symbol for offered traffic: the average number of attempted packet transmissions per packet transmission time.

GMSK Gaussian Minimum Shift Keying: a modification to frequency shift keying with a low modulation index.

GPS Geographic Positioning System: a U.S. government, satellite-based replacement for World War II techniques such as LORAN.

Hz Hertz: the term for cycles per second adopted to honor Heinrich Hertz.

LEAA Law Enforcement Assistance Administration: the funding agent that stimulated the first commercial uses of data radio technology.

LSA Licensed Space Arrangement.

MDC The 4800-bps protocol first deployed by Motorola in 1981 and the core of ARDIS.

MDI Mobile Data International: a Richmond, B.C., Canada, firm. First (1982) to employ Reed-Solomon coding techniques in commercial devices, their most successful customer was Federal Express. Purchased by Motorola, the company was gradually folded in as a division and now no longer exists as an independent entity.

MHX Mobitex Higher (Main) Exchange.

MIRS Motorola Integrated Radio System: an all-digital, high-speed (64 Kbps) system for voice and data (including paging, facsimile, etc.). The first user is Nextel, which brought Los Angeles live in August 1993.

MMP The 4800-bps protocol first deployed by MDI in 1982 and now being phased out.

MNP Microcom Networking Protocol: an error-detection/ retransmission scheme originally developed for wireline modems. A numbered series of increasingly rich offerings, it includes both data compression and cellular alternatives.

MOX Mobitex Area Exchange.

MRNE Mobile Radio New England: a street-level vehicular system.

MSA Metropolitan Statistical Area: the broader area surrounding a given city. The New York MSA is not just Manhattan, but also embraces portions of Long Island, Westchester, and New Jersey counties as well.

MTSO Mobile Telephone Switching Office: the cellular control center that coordinates and controls the activities of the cell sites and interconnects the telephones with the wireline network.

NAK Negative Acknowledgment: this can be a deliberate signal that the message was received in error, or can be inferred by time-out.

NCC Mobitex Network Control Center.

NCP Motorola (ARDIS) Network Control Processor.

PCMCIA Personal Computer Memory Card International Association (not People Can't Memorize Complicated Industry Acronyms). The "credit card" space for memory cards, and modems, to name a few.

RAM The public airtime service that became operational in 1990 and is now half owned by Bell South Mobility. It employs Ericsson protocols and infrastructure.

RD-LAP Radio Data Link Access Procedure: the 19,200-bps Motorola protocol first deployed commercially by ARDIS in October 1992.

RBOC Regional Bell Operating Company.

RMUG RadioMail's User Group.

S The symbol for steady state network throughput, the average number of successful transmissions per packet transmission time.

SABRE American Airline's reservation system whose original heritage drew on the developments of the SAGE system.

SAGE Semi-Automatic Ground Environment: the U.S. Air Force Air Defense System developed in the 1950s. A pioneer in data transmission, it used data radio from airborne early warning aircraft.

SMR Special Mobile Radio: originally a voice dispatch system, SMR frequencies are host to ARDIS, CoveragePLUS, MRNE, RaCoNet, and others.

SNA IBM's System Network Architecture.

SSI Signal Strength Indicator.

TCM Trellis Coded Modulation: additional (redundant) signal points added to a transmission, with only specific patterns of allowable sequences. If an impairment occurs, the closest pattern is selected to correct errors without retransmission.

TRIB Transfer Rate of Information Bits: the number of information bits accepted and the total time needed to send all bits (sync, header, CRC) to get those info bits accepted.

TSR Terminate and Stay Resident: Programs that load into memory and can be invoked by a specific interruption process.

UBER Undetected Bit (sometimes block) Error Rate: the percent of errors that escape error correction or detection and pass into the system.

VAN Value-Added Network: wireline transmission facility such as Advantis, CompuServe, Geisco, Telenet, or Tymnet that is employed to connect private computing facilities to data radio switching centers.

WAN Wide Area Network.

WAM Wide-Area Mobile: as in WAM workers.

Xcvr Transceiver: a combination radio transmitter and receiver.

APPENDIX B

IBM extends the range of the computer to the man on the beat

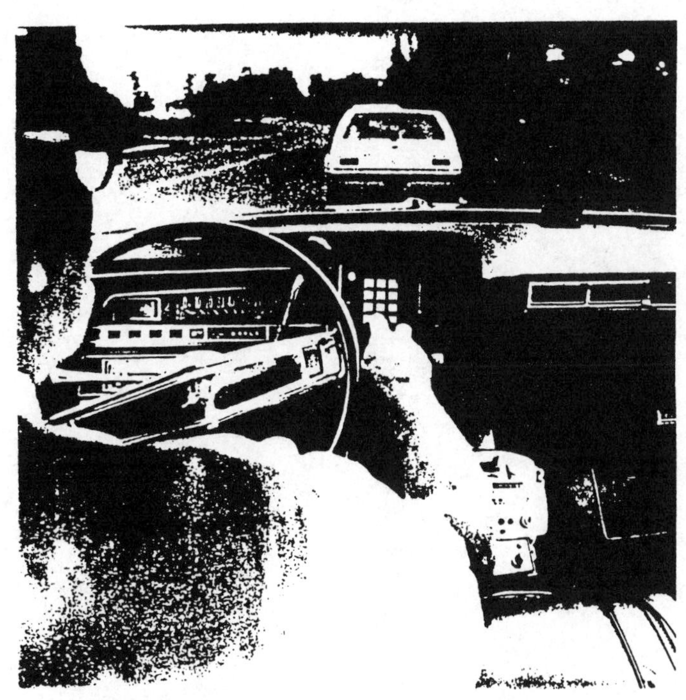

Photos-Simulated Application

Many law enforcement agencies use computer installations to provide centralized data banks of "auto alert" information on stolen or wanted vehicles. IBM can now extend the power of computer systems directly to the policeman in the patrol car, on motorcycle or walking a city beat. Using a new low-cost mobile/portable keyboard entry system developed by IBM, police can submit motor vehicle queries to a central computer via standard police radio channels. Status replies from the computer are communicated to the officer through audio-response computer-generated words.

Federal Systems Division

APPENDIX C

```
DIALING DIRECTORY: PCPLUS

     NAME                                      NUMBER   BAUD P D S D  SCRIPT
   1 NewsNet: TYMNET                         327-2974   1200 E-7-1 F
   2 NewsNet: TELENET                        359-9404   1200 E-7-1 F
   3 CompuServe Stamford                     964-1027   1200 E-7-1 F
   4 DataStorm BBS                   1 (314) 474-8477   1200 N-8-1 F
   5 Telenet Stamford                        359-9404   1200 N-7-1 F
   6 ibm boca                        1 (407) 443-3743   1200 E-7-1 F
   7                                                    1200 N-8-1 F
   8                                                    1200 N-8-1 F
   9                                                    1200 N-8-1 F
  10                                                    1200 N-8-1 F

     DIALING: CompuServe Stamford          LAST CONNECTED ON: 07/13/93
      NUMBER: 964-1027                  TOTAL COMPLETED CALLS: 302
 SCRIPT FILE:                           WAIT FOR CONNECTION: 45   SECS
   LAST CALL: CONNECT 1200               PAUSE BETWEEN CALLS: 4    SECS
 PASS NUMBER: 1                         TIME AT START OF DIAL: 09:12:04AM
ELAPSED TIME: 9                    TIME AT START OF THIS CALL: 09:12:05AM

Choice:    Space Recycle  Del Remove from list  End Change wait  Esc Abort
```

```
DIALING DIRECTORY: PCPLUS

     NAME                                     NUMBER   BAUD P D S D  SCRIPT
   1 NewsNet: TYMNET                        327-2974   1200 E-7-1 F
   2 NewsNet: TELENET                       359-9404   1200 E-7-1 F
   3 CompuServe Stamford                    964-1027   1200 E-7-1 F
   4 DataStorm BBS                1 (314) 474-8477   1200 N-8-1 F
   5 Telenet Stamford                       359-9404   1200 N-7-1 F
   6 ibm boca                     1 (407) 443-3743   1200 E-7-1 F
   7                                                   1200 N-8-1 F
   8                                                   1200 N-8-1 F
   9                                                   1200 N-8-1 F
  10                                                   1200 N-8-1 F

     DIALING: NewsNet: TELENET            LAST CONNECTED ON: 07/08/93
      NUMBER: 359-9404                TOTAL COMPLETED CALLS: 250
 SCRIPT FILE:                           WAIT FOR CONNECTION: 45  SECS
   LAST CALL: CONNECT 1200              PAUSE BETWEEN CALLS: 4   SECS
 PASS NUMBER: 1                       TIME AT START OF DIAL: 08:20:55AM
ELAPSED TIME: 13                 TIME AT START OF THIS CALL: 08:20:55AM

Choice:    Space Recycle  Del Remove from list  End Change wait  Esc Abort
```

```
PROCOMM PLUS on-line to NewsNet: TELENET at 1200 baud


TELENET
203 13C

TERMINAL=

@c net


NET CONNECTED
Welcome to NewsNet

Please sign on
-->id net22554
Password?



NET22554 (user 66) logged in Friday, 09 Jul 93 08:22:40.
Last login Thursday, 08 Jul 93 07:34:00.
```

```
                    ----------------
                    - N E W S N E T -
                    ----------------
             W O R K I N G   K N O W L E D G E

Copyright NewsNet 1982, 1993. Copying or redistribution by any means
strictly prohibited without prior written permission. For details, refer
to `Terms and Conditions of NewsNet Use' and the NewsNet Reference Guide,
or enter HELP TERMS at the main command prompt.

***New!  Market: Asia Pacific (IT47) presents and analyzes demographic
and lifestyle trends among consumers along the Pacific Rim.

***To find out what's new in business publications, databases, and research
techniques, access Business Information Alert (PB25).

***Phillips Business Information has recently added three publications. To
obtain a complete listing of all Phillips p
 Alt-Z FOR HELP| ANSI     | FDX |  1200 E71 | LOG CLOSED | PRINT OFF | ON-LINE
```

```
      Copyright (c) 1993
    CompuServe Incorporated
      All Rights Reserved

♣



 CompuServe                    TOP

  1 Access Basic Services
  2 Member Assistance (FREE)
  3 Communications/Bulletin Bds.
  4 News/Weather/Sports
  5 Travel
  6 The Electronic MALL/Shopping
  7 Money Matters/Markets
  8 Entertainment/Games
  9 Hobbies/Lifestyles/Education
 10 Reference
 11 Computers/Technology
 12 Business/Other Interests

 Enter choice number !
  Alt-Z FOR HELP| ANSI     | FDX |  1200 E71 | LOG CLOSED | PRINT OFF | ON-LINE
```

APPENDIX D

Table D-1 *Chicago: Motorola Heritage (852.4625)*

	Licensed		Cancelled		Cumulative		
User Name	**Date**	**Qty**	**Date**	**Qty**	**Gross**	**Cancel**	**Net**
Tandem	12-Nov-85	10			10	00	10
Honeywell	10-Dec-85	3			13	00	13
Xerox	10-Dec-85	37			50	00	50
Tandem			10-Mar-86	10	50	10	40
Sumer	15-Sep-86	2			52	10	42
Di-Namic Copy	20-Nov-86	3			55	10	45
Micro-Age Computer	20-Nov-86	3			58	10	48
Miller, Mason, Dickenson, Inc.	20-Dec-86	2			60	10	50
Motorola	31-Dec-86	122			182	10	172
Sun Micro Systems	09-Jan-87	5			187	10	177
Wallace Computer	09-Jan-87	3			190	10	180

Table D-1 *Chicago: Motorola Heritage (852.4625) (continued)*

	Licensed		Cancelled		Cumulative		
User Name	**Date**	**Qty**	**Date**	**Qty**	**Gross**	**Cancel**	**Net**
Ohlson Investigation	09-Jan-87	2			192	10	182
Micro-Age Computer			12-Jan-87	3	192	13	179
Miller, Mason, Dickenson, Inc.			12-Jan-87	2	192	15	177
Sorbus	10-Feb-87	16			208	15	193
Di-Namic Copy			13-Feb-87	3	208	18	190
Wallace Computer			13-Feb-87	3	208	21	187
Ohlson Investigation			13-Feb-87	2	208	23	185
Sun Micro Systems			20-Feb-87	5	208	28	180
National Deaf Society	24-Feb-87	5			213	28	185
Chicago Transit Authority	18-Mar-87	6			219	28	191
National Deaf Society			26-Mar-87	5	219	33	186
Chicago Research & Trading	24-Apr-87	4			223	33	190
BTI Midwest	08-Jun-87	10			233	33	200
Sumer			09-Jun-87	2	233	35	198
Tandem (7 net new)	22-Jun-87	17			250	35	215
Databroker Systems	19-Nov-87	2			252	35	217
Nielsen Market Research	23-Nov-87	2			254	35	219
Xerox			05-Jan-88	37	254	72	182
Quotron	24-May-88	1			255	72	183

Table D-1 *Chicago: Motorola Heritage (852.4625) (continued)*

	Licensed		Cancelled		Cumulative		
User Name	**Date**	**Qty**	**Date**	**Qty**	**Gross**	**Cancel**	**Net**
Baxter Healthcare	07-Sep-88	2			257	72	185
INET	18-Oct-88	3			260	72	188
CE Associates	16-Nov-88	3			263	72	191
Illinois Bell	29-Nov-88	2			265	72	193
AMA	14-Dec-88	3			268	72	196
Metromedia Paging	06-Jan-89	1			269	72	197
Peoples Gas, Coke & Light	15-Mar-89	4			273	72	201
United Airlines	19-Apr-89	5			278	72	206
Otis Elevator	31-May-89	15			293	72	221
White Way Sign	31-May-89	2			295	72	223
Tandem (additional)	15-Jun-89	26			321	72	249
Peoples Gas, Coke & Light			23-Jun-89	4	321	76	245
AIC Security	26-Jun-89	5			326	76	250
Motorola			27-Jun-89	7	326	83	243
Peoples Gas, Coke & Light	14-Dec-89	6			332	83	249
Charles River Computer	13-Mar-90	7			339	83	256
NCR	05-Jun-90	56			395	83	312
WW Grainger	24-Jul-90	10			405	83	322
Dataserv, Inc.	29-Mar-91	1			406	83	323
Avis Rent-a-Car	30-Dec-91	12			418	83	335
AT&T Paradyne	15-May-92	25			443	83	360

Table D-2 *Los Angeles: Motorola Heritage (851.7875)*

	Licensed		Cancelled		Cumulative		
User Name	**Date**	**Qty**	**Date**	**Qty**	**Gross**	**Cancel**	**Net**
Long Beach Container Terminal	26-Feb-87	8			8	00	8
Motorola C&E	12-Mar-87	63			71	00	71
Red Arrow Messenger	18-June-87	17			88	00	88
Westinghouse Beverage Group	19-Nov-87	20			108	00	108
Xerox	20-Nov-87	70			178	00	178
JAD Co.	27-Jun-87	2			180	00	180
City of Los Angeles	23-Aug-88	8			188	00	188
Stay Green, Inc.	07-Oct-88	4			192	00	192
Dial One AA Johnson Plumbing	07-Oct-88	4			196	00	196
Coast Club Service	28-Oct-88	5			201	00	201
Maersk Container	16-Nov-88	8			209	00	209
Wagstrom-Cartwright Computer	25-Nov-88	5			214	00	214
Metromedia Paging	06-Jan-89	1			215	00	215
Avicom International	09-Jan-89	5			220	00	220
California Financial Search	31-Mar-89	10			230	00	230
Citicorp TTI	31-Mar-89	5			235	00	235
California Financial Search			20-Apr-89	10	235	10	225
Otis Elevator	26-May-89	15			250	10	240
Tandem Computers	09-June-89	45			295	10	285

Table D-2 *Los Angeles: Motorola Heritage (851.7875) (continued)*

	Licensed		Cancelled		Cumulative		
User Name	**Date**	**Qty**	**Date**	**Qty**	**Gross**	**Cancel**	**Net**
ADAQ Systems Corp.	21-Jun-89	5			300	10	290
West Coast Express	30-Jun-89	4			304	10	294
PARSEC #1	12-Jul-89	15			319	10	309
City of Hermosa Beach	25-Jul-89	12			331	10	321
Matson Navigation	27-Nov-89	10			341	10	331
Movie Express	23-Jan-90	3			344	10	334
NCR	17-Apr-90	72			416	10	406
CCC Development	24-Jul-90	2			418	10	408
United Parcel Service	08-Aug-90	250			668	10	658
Dataserv, Inc.	02-Apr-91	3			671	10	661
Coca-Cola	08-Jan-92	68			739	10	729
AT&T Paradyne	05-May-92	35			774	10	764
Fleet Maintenance	12-Jul-92	110			884	10	874

Table D-3 *Manhattan: Motorola Heritage (851.2375)*

	Licensed		Cancelled		Cumulative		
User Name	**Date**	**Qty**	**Date**	**Qty**	**Gross**	**Cancel**	**Net**
City of New York	27-Sep-88	28			28	00	28
INET	18-Oct-88	1			29	00	29
QED Ringmaster	31-Oct-88	1			30	00	30
Alpha Page	16-Nov-88	2			32	00	32
Group Health, Inc.	16-Nov-88	5			37	00	37
Metromedia Paging Service	06-Jan-89	2			39	00	39

Table D-3 *Manhattan: Motorola Heritage (851.2375) (continued)*

	Licensed		Cancelled		Cumulative		
User Name	**Date**	**Qty**	**Date**	**Qty**	**Gross**	**Cancel**	**Net**
Port Authority of NY/NJ	03-Mar-89	2			41	00	41
National Valet Service	21-Mar-89	3			44	00	44
CBS, Inc.	24-Apr-89	3			47	00	47
Tandem Computers	09-Jun-89	45			92	00	92
Motorola, Inc.	11-Jul-89	56			148	00	148
City of Newark	03-Aug-89	40			188	00	188
SIE, Inc.	12-Sep-89	3			191	00	191
NYC Marshal Airday	30-May-90	2			193	00	193
NCR	05-Jun-90	25			218	00	218
NYC Marshal Burko	02-Aug-90	1			219	00	219
Carolyn O'Connor	07-Aug-90	2			221	00	221
United Parcel Service	08-Aug-90	225			446	00	446
Dow Jones	14-Sep-90	5			451	00	451
NYC Transportation Dept.	28-Nov-90	60			511	00	511
Dataserv, Inc.	29-Mar-91	3			514	00	514
Guaranteed Overnight Delivery	15-Apr-91	25			539	00	539
NCR (additional)	15-Apr-91	44			583	00	583
Avis Rent-a-Car	02-Aug-91	35			618	00	618
AT&T Paradyne	03-Dec-91	15			633	00	633
Jacob Sampson	20-Mar-92	1			634	00	634

APPENDIX E

Table E-1 *Greater New York City*

User Name	WNIX499		WNKM884		Cumulative Total
	Licensed	Qty	Licensed	Qty	
MB Trucking	03-Jul-91	17	23-Jul-91	17	34
ConRail	07-Feb-92	33	07-Feb-92	33	100
American Courier Express	27-Feb-92	1	27-Feb-92	1	102
Hoboken Fire Dept.	02-Mar-92	2	02-Mar-92	2	106
Regional Comms., Inc.	02-Mar-92	1	02-Mar-92	1	108
Ericsson/GE Mobile Data	04-Mar-92	15.62	04-Mar-92	15.62	139.24
RAM Mobile Data	11-Mar-92	25	11-Mar-92	25	189.24
National Car Rental	22-Apr-92	6	22-Apr-92	8	203.24
British Airways	06-Jul-92	2	06-Jul-92	2	207.24
Travelers Insurance	17-Jul-92	3	17-Jul-92	3	213.24
Dow Jones Field Service	30-Jul-92	34	30-Jul-92	33	280.24
PMI Food Equipment Group	31-Jul-92	0.5	31-Jul-92	0.5	281.24

Table E-1 *Greater New York City (continued)*

User Name	WNIX499		WNKM884		Cumulative Total
	Licensed	Qty	Licensed	Qty	
Fleet Maintenance, Inc.	04-Aug-92	0.5	04-Aug-92	0.5	282.24
Performance Systems Intl.	05-Aug-92	5	05-Aug-92	4	291.24
Anterior Technology	10-Sep-92	4	10-Sep-92	4	299.24

Table E-2 *Greater Chicago*

User Name	WNID259		WNKL297		Cumulative Total
	Licensed	Qty	Licensed	Qty	
Ericsson/GE Mobile Data			04-Mar-92	15.62	15.62
RAM Mobile Data			11-Mar-92	11	26.62
National Car Rental			22-Apr-92	4	30.62
Chicago Parking Authority			26-Jun-92	4	34.62
British Airways			06-Jul-92	2	36.62
Dow Jones Service			30-Jul-92	10	46.62
Fleet Maintenance, Inc.			04-Aug-92	1	47.62
Performance Systems, Intl.			05-Aug-92	4	51.62
Anterior Technology			10-Sep-92	4	55.62

Table E-3 *Orange and Los Angeles Counties*

User Name	WNJA872		WNJA904		Cumulative Total
	Licensed	Qty	Licensed	Qty	
Ericsson/GE Mobile Data	04-Mar-92	15.62	04-Mar-92	15.62	31.24
RAM Mobile Data	11-Mar-92	3	11-Mar-92	3	37.24
National Car Rental	22-Apr-92	3	22-Apr-92	4	44.24
Coast Shuttle, Inc.	28-May-92	14			58.24
United Way	05-Jun-92	1			59.24
Ideal Container Corp.	30-Jun-92	2			61.24
British Airways	06-Jul-92	2			63.24
Dow Jones Service	30-Jul-92	7	30-Jul-92	7	77.24
PMI Food Equipment	31-Jul-92	0.5			77.74
Sun Micro Systems	03-Aug-92	12	03-Aug-92	12	101.74
Fleet Maintenance, Inc.	04-Aug-92	0.5			102.24
Performance Systems, Intl.	05-Aug-92	4			106.24
Anterior Technology	10-Sep-92	4			110.24

APPENDIX F

User	Omni-tracs	Contract value($)	Install				
			Begin	End	Host	Source	Date
Schneider National Van	5000	20,000,000	1988	Aug-89	"mainframe"	SN	05-Dec-88
Munson Transportation	1000	4,000,000				MDR	15-Feb-89
Direct Transit	1300	5,000,000				IC	27-Oct-89
R&D Trucking	200					IC	27-Oct-89
Leaseway Technology	200	800,000				IC	27-Oct-89
Waggoner's Trucking	200					IC	27-Oct-89
Boyds Brothers	300					IC	02-Mar-90
Roberts Express	1150			Dec-90		MDR	07-May-90
Star Delivery	250	1,000,000				MSR	Jun-90
Northern Telecomm	40					MDR	18-Jun-90
Arrow Trucking	250	1,000,000				MSR	16-Nov-90
MNX	700	3,000,000				MDR	03-Dec-90
Thermo King	400		Dec-90			MSR	14-Dec-90
TRM Starcom	35			Oct-91	NCR 7000	ERT	05-Feb-92

Table F-1 *Omnitracs Users*

User	Omni-tracs	Contract value($)	Install				
			Begin	End	Host	Source	Date
U.S. Xpress	800	5,300,000			IBM AS/400	MSN	Dec-91
Victory Express	400		1991	Mar-92		ERT	18-Mar-92
Buske Lines	1000	4,000,000				ERT	18-Mar-92
Continental Express							
Creech Brothers Truck							
Pryslak Trucking							
Rich Refrigerated Intl.							
J.B. Hunt	6000		May-92		IBM 3090	ERT	15-Apr-92
KLLM Transport					IBM AS/400	ERT	15-Apr-92
Star Delivery						ERT	15-Apr-92
Interstate Distributor					IBM AS/400	ERT	15-Apr-92
Navajo Express	675	4,300,000	Apr-92		IBM AS/400	ERT	29-Apr-92
Maverick Transportation	200		Feb-92	May-92	IBM AS/400	ERT	05-Aug-92

Table F-1 *Omnitracs Users (continued)*

User	Omni-tracs	Contract value($)	Install				
			Begin	End	Host	Source	Date
Digby Truck Lines							
Goodway							
Landair Transport	1400	8,500,000				MSR	14-Sept-92
Dick Simon Trucking							
Trans State Lines							
Wiley Sanders							
Werner Enterprises	2700	12,000,000				MSR	12-Oct-92
Freymiller Trucking	700	3,000,000				MSR	09-Nov-92
C.R. England & Sons	1150				IBM AS/400	MSR	17-Feb-92
Appalachian Freight							
Daymark							
Donco							
Eaglewings Systems	1300					MSR	21-Dec-92
Foreway Transport							
RTC							

Table F-1 *Omnitracs Users (continued)*

User	Omni-tracs	Contract value($)	Install		Host	Source	Date
			Begin	End			
Prime	850	4,000,000				ERT	21-Dec-92
CCC Express		14,000,000				TlcmR	29-Mar-93
North American Van Lines							
Jevic Transportation							
Westway Express							
Marten Transport	735	3,000,000				TlcmR	05-Apr-93
Buford TV (Tyler,,TX)	70					ComD	10-Jun-93
Total	29005						

Table F-1 *Omnitracs Users (continued)*

APPENDIX G

Number of Home Health Agencies in the United States

State	Total 1991	Total 1986	% Change	State	Total 1991	Total 1986	% Change
Alabama	168	111	51.4	Montana	46	20	130.0
Alaska	10	3	233.3	Nebraska	103	30	243.3
Arizona	130	70	85.7	Nevada	37	16	131.3
Arkansas	195	183	6.6	New Hampshire	72	42	71.4
California	644	401	60.6	New Jersey	143	54	164.8
Colorado	142	109	30.3	New Mexico	70	42	66.7
Connecticut	152	123	23.6	New York	486	144	237.5
Delaware	33	21	57.1	North Carolina	275	107	157.0
Dist. Col.	19	14	35.7	North Dakota	40	24	66.7
Florida	619	148	318.2	Ohio	357	204	75.0
Georgia	151	74	104.1	Oklahoma	157	123	27.6
Hawaii	26	0	---	Oregon	85	60	41.7
Idaho	32	23	39.1	Pennsylvania	360	199	80.9
Illinois	440	288	52.8	Puerto Rico	40	33	21.2
Indiana	211	106	99.1	Rhode Island	24	15	60.0
Iowa	187	126	48.4	South Carolina	75	36	108.3
Kansas	207	112	84.8	South Dakota	37	17	117.7
Kentucky	162	95	70.5	Tennessee	409	253	61.7
Louisiana	263	119	121.0	Texas	811	490	65.5
Maine	56	20	180.0	Utah	62	22	181.8
Maryland	106	96	10.4	Vermont	24	21	14.3
Massachusetts	199	161	23.6	Virginia	186	95	95.8
Michigan	229	157	45.9	Washington	150	42	257.1
Minnesota	220	143	53.9	West Virginia	65	33	97.0
Mississippi	121	135	-10.4	Wisconsin	187	118	58.5
Missouri	224	175	28.0	Wyoming	38	30	26.7
				Total U.S.	**9,285**	**5,283**	**75.8**

Data Source: SMG Marketing Group Inc. Copyright 1992

Home Health Care Financials

Average Annual Revenues and Expenses, by Size of Agency

Revenues/Expenses	Number of Patient Visits per Week 1-100	101-300	301-500	501+	Industry Avg.
Total Patient Revenues	$822,405	$1,769,239	$2,287,664	$5,800,718	$2,790,625
Less: Allowances and Discounts	$59,298	$199,803	$355,581	$1,308,986	$509,020
Net Patient Revenues	$763,107	$1,569,436	$1,932,083	$4,491,732	$2,281,605
Total Operating Expenses	$718,827	$1,443,013	$1,713,255	$3,885,998	$2,020,784
Net Income From Services	$44,280	$126,423	$218,828	$605,734	$260,821
Total Other Income*	$7,739	$18,921	$10,277	$37,204	$19,765
Pre-tax Profit (Loss)	$52,019	$145,344	$229,105	$642,938	$280,586

** Other includes income from the sale and rental of supplies and equipment, fund raising, grants, charitable contributions, local government funding, investments and interest.*

Data Source: SMG Marketing Group Inc. Copyright 1992

Home Care Statistics

Number of Medicare beneficiaries receiving home care:	2.1 million (1990)
Number of total home health visits under Medicare:	54.8 million (1990)
Average cost per visit:	$66.00 (1990)
Total Medicare spending on home health care:	$3.5 billion (1990)
Percent of Medicare budget spent on home health care:	3.3% (1990)
Total Medicaid spending on home health care:	$3.4 billion (1990)
Percent of Medicaid budget spent on home care:	5.2% (1990)
Total spending for home care by all payers:	$16.2 billion (1990)
Percent of national health care spending for home care by all payers:	2.4% (1990)

Statistics provided by National Association for Home Care study

% Increase in Medicare Costs (1985-1990)

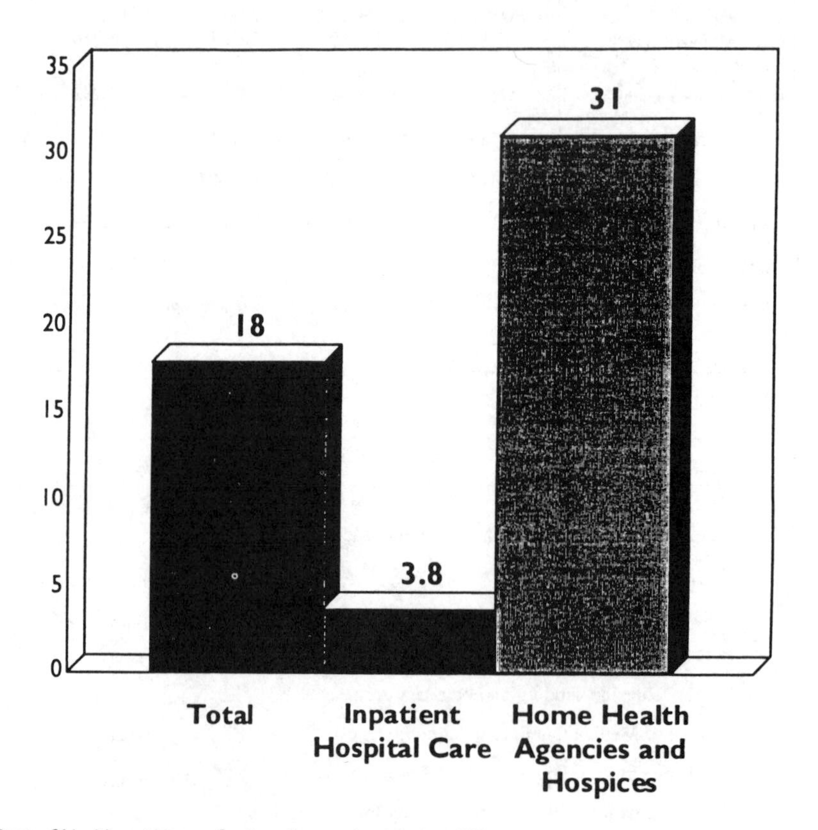

Source: Dept. of Health and Human Services; Congressional Budget Office

4365 Route One, Princeton, NJ 08540-5705 • 609/987-8181 • FAX: 609/987-8874
Regional Offices: Oakland, CA. • Pompano Beach, FL • Downers Grove, IL

082692DAW

APPENDIX H

MobiData
Communications

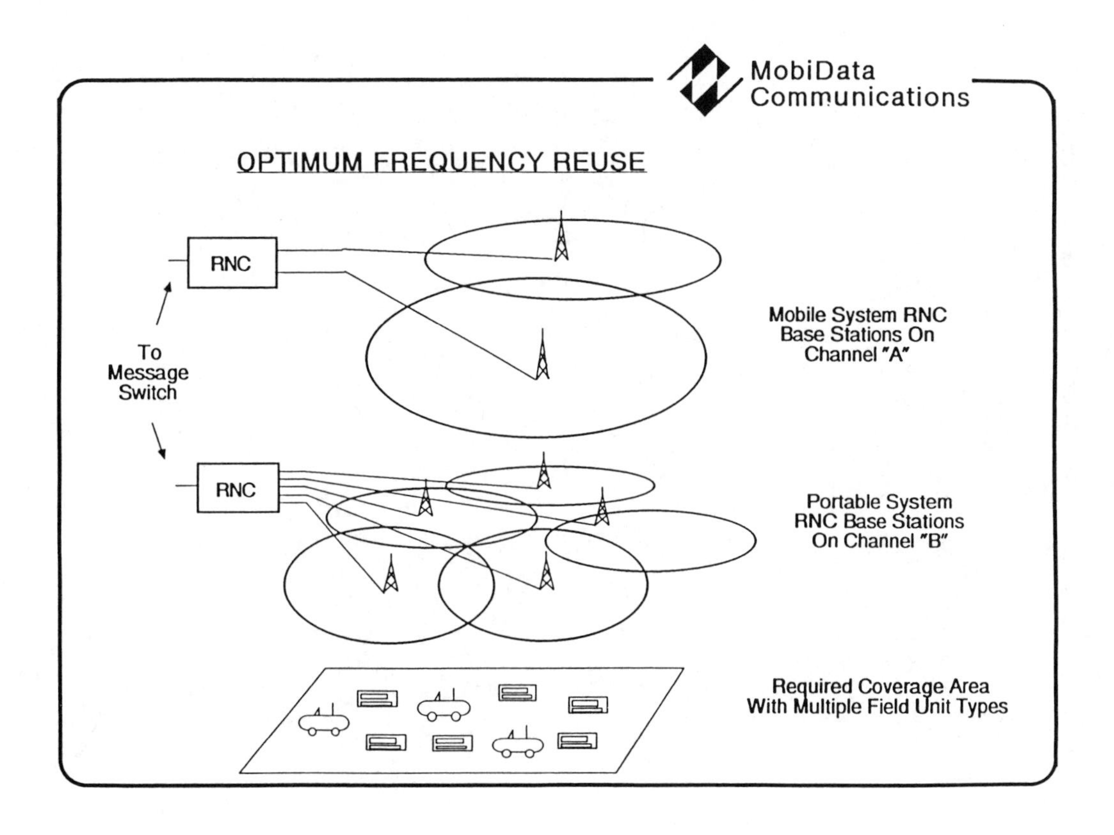
OPTIMUM FREQUENCY REUSE
RNC
To
Message
Switch
RNC
Mobile System RNC
Base Stations On
Channel "A"
Portable System
RNC Base Stations
On Channel "B"
Required Coverage Area
With Multiple Field Unit Types

APPENDIX I

Field Service Cost/Saving Analysis Sample

	LA	PORT	DTR	BOS	TOTAL
Savings:					
Number of Service Calls/Year	70,460	26,936	19,136	41,756	158,288
Number of FE Calls to Dispatch/Year	100,758	56,027	41,334	40,921	239,039
Dispatcher Cost Savings:					
Handling FE Phone Calls	82,218	40,788	25,875	27,171	176,052
Jotting Call Close Data				6,253	6,253
Dispatcher Sub-Total:	82,218	40,788	25,875	33,424	**$182,305**
Field Technician Cost Savings:					
Find Phone/Access	53,687	15,463	9,982	11,765	90,897
Closing Calls Over Phone	70,765	35,829	17,458	19,528	143,581
Opening Calls Over Phone	66,984	23,852	12,837	19,368	123,041
Hold Time on Phone	126,292	45,104	33,506	82,476	287,378
Field Technician Sub-Total:	317,728	120,249	73,784	133,137	**$644,897**
Telephone Operations Cost Savings:					
Dispatch Time Spent on Phone	35,366	26,512	17,661	17,661	96,358
FE Hold Time on Phone	42,829	15,296	27,970	27,970	97,459
Telephone Sub-Total:	78,195	41,808	45,632	45,632	**$193,817**
Paging Cost Savings:					
Pager Cost	5,292			4,608	**$9,900**
Total Projected Savings:	483,434	202,844	127,840	216,800	**$1,030,919**
Cost:					
ARDIS Service	117,600	42,000	31,200	76,800	267,600
RPM-840C's	392,000	140,000	104,000	256,000	892,000
RPM-840C Maintenance	17,640	6,300	4,680	11,520	40,140
Integration Services	98,000	35,000	26,000	64,000	223,000
Total Projected Cost	625,240	223,300	165,880	408,320	**$1,422,740**
Investment Payback (Years):					**1.38**

Real-time information solutions
for real-life business problems

ARDIS

APPENDIX J

```
CompuServe Information Service
 13:49 EDT Monday 09-Aug-93 P

 Last access: 08:47 29-Jul-93

     Copyright (c) 1993
   CompuServe Incorporated
     All Rights Reserved

You have Electronic Mail waiting.
GO RATES for current information

What's New This Week

 1 Forum Opens for Newton Talk
 2 Travel Britain Online Debuts
 3 Free Articles in Health Database
 4 Three IQuest SmartSCANs on Sale
 5 Stay Informed with Global Report
 6 Student's Forum Hosts Europe CO
 7 Pacific Forum Celebrates Birthday
 8 Fenics II Network Available in Japan
 9 Win Plane, Game Tickets in The Mall
   (Above Articles Are Free)
10 Online Today
11 Special Events/Contests Area (Free)

Enter choice !go mail

CompuServe Mail  Main Menu

 1 READ mail, 1 message pending
 2 COMPOSE a new message
 3 UPLOAD a message
 4 USE a file from PER area
 5 ADDRESS Book
 6 SET options
 9 Send a CONGRESSgram ($)

Enter choice !1
```

```
CompuServe Mail

Date:  31-Jul-93 12:17 EDT
From:  INTERNET:jcb@vnet.IBM.COM
Subj:  wedding

Sender: jcb@vnet.IBM.COM
Received: from vnet.ibm.com by iha.compuserve.com (5.67/5.930129sam)
        id AA02997; Sat, 31 Jul 93 12:17:28 -0400
Message-Id: <9307311617.AA02997@iha.compuserve.com>
Received: from LGEVM2 by vnet.IBM.COM (IBM VM SMTP V2R2) with BSMTP id 6484;
   Sat, 31 Jul 93 12:15:56 EDT
Date: Sat, 31 Jul 93 18:08:14 FST
From: jcb@vnet.IBM.COM
To: 73260.2036@compuserve.com
Subject: wedding

Hi Jim, did you receive our response to your note?


Last page.  Enter command or <CR> to continue !

CompuServe Mail  Action Menu

** INTERNET:jcb@vnet.IBM.COM/wedding **

 1 DELETE this message
 2 FILE in PER area
 3 FORWARD
 4 REREAD message
 5 REPLY
 6 SAVE in mailbox
 7 DOWNLOAD message


Enter choice or <CR> to continue !1

CompuServe Mail  Main Menu

   *** No mail waiting ***

 2 COMPOSE a new message
 3 UPLOAD a message
 4 USE a file from PER area

 5 ADDRESS Book
 6 SET options

 9 Send a CONGRESSgram ($)


Enter choice !off

Thank you for using CompuServe!

Off at 13:49 EDT 9-Aug-93
Connect time = 0:01


Host Name:
```

APPENDIX K

This simulation model tests a 19-channel cell employing the Motorola "round robin" voice-channel assignment technique. The traffic philosophy is Erlang[B]: blocked calls are cleared; there is no camp-on. The program tests for violation of 2 percent blocking at data entry time. The user is warned that the parameters of choice may cause problems. However, it is possible to proceed and deliberately overdrive the cell. After priming the cell, the measured interval is one hour at one-second granularity.

KEY ASSUMPTIONS

1. The random length of each call is calculated within a range from zero to twice the length of the mean call entered at RUN time. That is, if a mean call length of 120

seconds is chosen, calls will range from 0 to 240 seconds. In this example there is no provision for a call that is, say, 10 minutes long.

2. Only one data channel is sought from the collected pool of voice channels that are temporarily free.
3. The data pointer moves preemptively in an effort to stay just behind the continuously progressing voice pointer. Assume the voice pointer is on channel 13, the next channel to be assigned a voice call, and the data pointer is on channel 5. If, say, channel 11 is freed by the termination of a voice call, the data pointer will be moved to channel 11. It will never regress to a less favorable position.

ENTERING DATA

1. When the model is loaded the user is prompted for the number of calls/hour that are to be tested. After entry the model responds with the average call rate in seconds.
2. The user is then prompted for the mean call length in seconds. The model responds with the testable time range of the call.
3. Because it is unrealistic to assume each call is an independent event, the model permits the user to shape peak periods from the keyboard. Initially, assume no peak period testing is desired.
    a. The user is first asked the rate at which voice calls should be initiated during an intensity peak. If "1" is entered, no peak period will exist. Every call will be, on average, probabilistically attempted at the average rate.
    b. The user will then be prompted for the length of the peak period. Enter "0" because no intensity peaks are desired.
    c. Similarly, the user will also enter "0" for the number of peak periods.

In this nonpeak case the model will summarize the results and begin execution.

4. It is highly likely that the effect of intensity peaks will be of interest to the user. In this case the model prompts are illustrated as follows:
   a. The average rate multiplier can be any value that does not exceed one voice call per second, the model granularity.
   b. The length of the peak period specifies how long the peak message rate should be sustained.
   c. The number of peak periods specifies how often the user wants the peak period to occur during the one-hour measured run. This number is randomly derived during the model execution. The user may specify one period but get none; it is also routine (in this example) to get two periods per hour.

The model summary in the intensity case is instructive. The model will attempt to execute the peak rate during the peak period. However, there will be some shortfall between the rate attempted and the rate achieved.

Note, also, that the model attempts to hold the one-hour rate at the original Erlang level. If peak periods are specified, the non-peak periods will see call rate reductions.

5. Input modifications:
   a. If the initial message rate times length calculations exceed 2 percent blocking probability, the model flags the user of this fact. However, the user is not constrained to modify the input; the model will proceed after issuing a warning.
   b. If the intensity peak is too bizarre (in the author's judgment), the model requires the user to modify the input to a more realistic scenario.

```
Running

     .---------------------------------------------------------.
     |   hoptest7   JFD Associates    copyright: 2-16-92        |
     |                                                          |
     |  Simulation of a 19 channel cell with 2% blocking        |
     |  probability (Erlangs[B]=12.3) over one peak hour        |
     |  at one second granularity.                              |
     '---------------------------------------------------------'

FROM THE KEYBOARD:

   enter the average number of voice calls per hour:
```



```
Running

     .---------------------------------------------------------.
     |   hoptest7   JFD Associates    copyright: 2-16-92        |
     |                                                          |
     |  Simulation of a 19 channel cell with 2% blocking        |
     |  probability (Erlangs[B]=12.3) over one peak hour        |
     |  at one second granularity.                              |
     '---------------------------------------------------------'

FROM THE KEYBOARD:
```



```
   enter the average number of voice calls per hour: 369
   this is an average of one call every 9.8 seconds.

   enter the mean voice call duration in seconds:
```

Running

```
.-----------------------------------------------------.
|   hoptest7   JFD Associates   copyright: 2-16-92    |
|                                                     |
|  Simulation of a 19 channel cell with 2% blocking   |
|  probability (Erlangs[B]=12.3) over one peak hour   |
|  at one second granularity.                         |
`-----------------------------------------------------'
```

FROM THE KEYBOARD:

```
enter the average number of voice calls per hour: 369
this is an average of one call every 9.8 seconds.

enter the mean voice call duration in seconds: 120
the call length will vary from  0 to  240 seconds.

create intensity peaks by:

1 - selecting an average rate multiplier:
```

Running

```
.-----------------------------------------------------.
|   hoptest7   JFD Associates   copyright: 2-16-92    |
|                                                     |
|  Simulation of a 19 channel cell with 2% blocking   |
|  probability (Erlangs[B]=12.3) over one peak hour   |
|  at one second granularity.                         |
`-----------------------------------------------------'
```

FROM THE KEYBOARD:

```
enter the average number of voice calls per hour: 369
this is an average of one call every 9.8 seconds.

enter the mean voice call duration in seconds: 120
the call length will vary from  0 to  240 seconds.

create intensity peaks by:

1 - selecting an average rate multiplier: 1
    the peak rate will be one call every 9.8 seconds

2 - selecting length of a peak period in seconds:
```

```
.--------------------------------------------------------.
|  hoptest7   JFD Associates    copyright: 2-16-92       |
|                                                        |
|  Simulation of a 19 channel cell with 2% blocking      |
|  probability (Erlangs[B]=12.3) over one peak hour      |
|  at one second granularity.                            |
`--------------------------------------------------------'
```

FROM THE KEYBOARD:

enter the average number of voice calls per hour: 369
this is an average of one call every 9.8 seconds.

enter the mean voice call duration in seconds: 120
the call length will vary from 0 to 240 seconds.

create intensity peaks by:

1 - selecting an average rate multiplier: 1
    the peak rate will be one call every 9.8 seconds

2 - selecting length of a peak period in seconds: 0

3 - selecting the number of peak periods desired:


```
|  at one second granularity.                            |
`--------------------------------------------------------'
```

FROM THE KEYBOARD:

enter the average number of voice calls per hour: 369
this is an average of one call every 9.8 seconds.

enter the mean voice call duration in seconds: 120
the call length will vary from 0 to 240 seconds.

create intensity peaks by:

1 - selecting an average rate multiplier: 1
    the peak rate will be one call every 9.8 seconds

2 - selecting length of a peak period in seconds: 0

3 - selecting the number of peak periods desired: 0

```
during    0     peak seconds ~   0 calls will be attempted.
during 3600 non-peak seconds ~369 calls will be attempted.
non-peak call average arrival rate is one every 9.8 seconds.
```

Running

```
 .-----------------------------------------------------.
 |  hoptest7   JFD Associates   copyright: 2-16-92   |
 |                                                     |
 | Simulation of a 19 channel cell with 2% blocking   |
 | probability (Erlangs[B]=12.3) over one peak hour   |
 | at one second granularity.                         |
 `-----------------------------------------------------'

FROM THE KEYBOARD:

   enter the average number of voice calls per hour: 387
   this is an average of one call every 9.3 seconds.

   enter the mean voice call duration in seconds: 120
   the call length will vary from  0 to  240 seconds.

NOTE: your parameters exceed the 2% blocking probability.
      Do you want to change them?  y/n.
n
Proceeding.  Expect additional blocked calls.
   create intensity peaks by:

   1 - selecting an average rate multiplier: 9.3
```


```
 |  hoptest7   JFD Associates   copyright: 2-16-92   |
 |                                                     |
 | Simulation of a 19 channel cell with 2% blocking   |
 | probability (Erlangs[B]=12.3) over one peak hour   |
 | at one second granularity.                         |
 `-----------------------------------------------------'

FROM THE KEYBOARD:

   enter the average number of voice calls per hour: 387
   this is an average of one call every 9.3 seconds.

   enter the mean voice call duration in seconds: 120
   the call length will vary from  0 to  240 seconds.

NOTE: your parameters exceed the 2% blocking probability.
      Do you want to change them?  y/n.
n
Proceeding.  Expect additional blocked calls.
   create intensity peaks by:

   1 - selecting an average rate multiplier: 9.3
```


```
       the peak rate will be one call every 1.0 seconds

   2 - selecting length of a peak period in seconds: 60
```

```
            | Simulation of a 19 channel cell with 2% blocking  |
            | probability (Erlangs[B]=12.3) over one peak hour  |
            | at one second granularity.                        |
            -----------------------------------------------------

FROM THE KEYBOARD:

   enter the average number of voice calls per hour: 387
   this is an average of one call every 9.3 seconds.

   enter the mean voice call duration in seconds: 120
   the call length will vary from  0 to  240 seconds.

NOTE: your parameters exceed the 2% blocking probability.
      Do you want to change them?  y/n.
n
Proceeding.  Expect additional blocked calls.
   create intensity peaks by:

   1 - selecting an average rate multiplier: 9.3
       the peak rate will be one call every 1.0 seconds

   2 - selecting length of a peak period in seconds: 60

   3 - selecting the number of peak periods desired: 1



   enter the average number of voice calls per hour: 387
   this is an average of one call every 9.3 seconds.

   enter the mean voice call duration in seconds: 120
   the call length will vary from  0 to  240 seconds.

NOTE: your parameters exceed the 2% blocking probability.
      Do you want to change them?  y/n.
n
Proceeding.  Expect additional blocked calls.
   create intensity peaks by:

   1 - selecting an average rate multiplier: 9.3
       the peak rate will be one call every 1.0 seconds

   2 - selecting length of a peak period in seconds: 60

   3 - selecting the number of peak periods desired: 1

   during   60      peak seconds ~ 60 calls will be attempted.
   during 3540 non-peak seconds ~327 calls will be attempted.
   non-peak call average arrival rate is one every 10.8 seconds.
```

```
Running

        .----------------------------------------------------.
        |  hoptest7   JFD Associates   copyright: 2-16-92     |
        |                                                     |
        | Simulation of a 19 channel cell with 2% blocking    |
        | probability (Erlangs[B]=12.3) over one peak hour    |
        | at one second granularity.                          |
        '-----------------------------------------------------'

FROM THE KEYBOARD:

   enter the average number of voice calls per hour: 387
   this is an average of one call every 9.3 seconds.

   enter the mean voice call duration in seconds: 120
   the call length will vary from  0 to  240 seconds.

NOTE: your parameters exceed the 2% blocking probability.
      Do you want to change them?  y/n.
```

```
Running

        .----------------------------------------------------.
        |  hoptest7   JFD Associates   copyright: 2-16-92     |
        |                                                     |
        | Simulation of a 19 channel cell with 2% blocking    |
        | probability (Erlangs[B]=12.3) over one peak hour    |
        | at one second granularity.                          |
        '-----------------------------------------------------'

FROM THE KEYBOARD:

   enter the average number of voice calls per hour: 387
   this is an average of one call every 9.3 seconds.

   enter the mean voice call duration in seconds: 120
   the call length will vary from  0 to  240 seconds.

NOTE: your parameters exceed the 2% blocking probability.
      Do you want to change them?  y/n.
n
Proceeding.  Expect additional blocked calls.
```

```
| at one second granularity.                          |
'------------------------------------------------------'

FROM THE KEYBOARD:

   enter the average number of voice calls per hour: 369
   this is an average of one call every 9.8 seconds.

   enter the mean voice call duration in seconds: 120
   the call length will vary from  0 to  240 seconds.

   create intensity peaks by:

   1 - selecting an average rate multiplier: 9
       the peak rate will be one call every 1.1 seconds

   2 - selecting length of a peak period in seconds: 600

   3 - selecting the number of peak periods desired: 1

   half the calls are occurring in the peak.  RE-ENTER

   create intensity peaks by:

   1 - selecting an average rate multiplier:
```

INTERPRETING OUTPUT

1. The first five lines printed are a comparison of parameters entered versus what the model actually achieved. It also includes an indicator of whether or not the simulated message rate and length exceeded the 2 percent blocking level.
   a. In the no-peak-period example the simulated message rate was ~3 percent higher than requested; the average message length was ~1.5 percent longer than requested. This combination yields ~12.8 Erlangs; 2 percent blocking is 12.3 Erlangs for 19 channels.
   b. In the intensive peak example the model rate/length variations yield ~12.36 Erlangs—a borderline case. The peak rate multiplier was set to attempt a new voice call every second of the peak period. The peak period itself was 60 seconds long, and only one peak period occurred during the simulated hour.
2. Priming statistics describe the state of the simulated cell at the start of the measured hour. Nineteen voice calls are started in sequence at the average message rate. Unless the rate is very high, some of these voice calls will have been ended by the time the 19th is initiated.
   a. In the first example the 19th call started in 164 seconds; simply playing the averages would lead one to expect ~171 seconds. By the time the 19th call started, eight calls had already ended.
   b. In example two, the 19th call was initiated rather quickly—134 seconds—thus only three calls had finished.

   Priming is used to ensure that a "going cell" is measured during the one-hour interval.
3. The central matrix summarizes the channel-by-channel activity during the measured hour. The distribution of calls ended is quite uniform: 17–23 in case 1; 17–24 in case 2.

Channels with calls in progress at the end of the measured hour have a "1" set under the active call column heading. The time remaining for each of those calls is listed under the "seconds to go" heading.

The average call duration per channel is listed under "seconds/call." Because each call in these examples could range from 0–240 seconds, the average call length tends to swing ±4 percent from the mean requested. A longer measured interval would tend to dampen those swings.

Channel utilization is the straightforward calculation of total seconds utilized for voice, divided by 3,600 seconds. The results are in line with the utilization expected at ~12.3 Erlangs.

4. The five-line summary following the channel statistics summarizes key incidents of interest:
   a. Voice calls blocked counts the number of times a voice call was attempted when all 19 channels were already in use. All busy calls are cleared; there is no camp-on.
   b. Total planned data hops counts the number of times in an hour that the data pointer was moved, following the preemptive strategy described earlier. The total 176, which appears in both examples, is coincidental, but the variation in this number tends to be modest for near-equivalent Erlang levels.
   c. The number of times all channels are busy is self-explanatory. But note that 19 busy channels blocks or aborts data messages, but does not necessarily cause much slower rate voice messages to be lost.
   d. Individual busy periods is a variation on "all channels busy" to account for those cases in which busy periods occur back-to-back, thus causing an extended, unbroken busy period.
   e. Cumulative busy periods is the sum of all seconds in which the 19 channels were busy.

The difference between no blocked periods and a single, high-intensity interval is obvious in these summary examples: voice calls blocked, expected to be ~8, jumps from 3 to 50; the times all channels busy doubled; and busy seconds climbs more than four times, thus doubling the average number of seconds per blocked interval.

5. The blocked second histogram details the key intervals in a more useful way. In the peak period example, the average blocked time is 7.4 seconds. One-third of the blocked events were four seconds or less. But more than a third were ten seconds or more, with the worse block double the average.

PROGRAM WEAKNESSES

1. The program operates with Erlang[B] philosophy: All busy calls are cleared. If Erlang[C] were to be used, modification to the table-checking routines could be inserted within an hour. But if busy calls camp-on, a new queue structure must be created, which requires about two days of new work. Erlang[C] results are expected to be more pessimistic for data during voice peak periods.
2. The average call length is inclusive of all setup and tear-down time. For example, when a voice call ends the model immediately releases the channel for new use. In practice, this does not happen. Transmitter guard band setup times may be several seconds in duration, blocking the channel from new work. Although the effect of setup/tear-down times can be approximated by lengthening the average call time, separate routines should be constructed in which these are user-controllable parameters.
3. The histogram records the end of a busy period if one free second is detected. In practice, blocked periods tend to occur in bursts with scattered one-second free periods. In this short span a data-hopping system would seek out the free channel and then get ready (or even commence) to transmit, only to have to abort. An alternative representation of the busy period should be developed.

	ENTERED	SIMULATED		>2% BLOCK
voice calls/hour:	387	362		
mean call length, seconds:	120	123	(average)	YES
arrival rate multiplier:	9.3	9.3		
peak period length (sec):	60	60		
number of peak periods:	1	1		

Priming Statistics:
elapsed time at start of 19th call: 134 seconds.
calls in process: 16

One Hour Run-Time Statistics:

channl	#calls ended	active calls	second to-go	second /call	total second	utili-zation
1	19	0	0	120	2283	0.63
2	17	1	23	142	2410	0.67
3	17	1	53	154	2618	0.73
4	20	0	0	109	2182	0.61
5	20	0	0	105	2091	0.58
6	20	1	157	121	2423	0.67
7	17	1	36	136	2311	0.64
8	19	1	84	126	2386	0.66
9	21	1	14	102	2137	0.59
10	20	0	0	112	2239	0.62
11	18	0	0	136	2454	0.68
12	18	1	41	134	2411	0.67
13	18	1	14	135	2433	0.68
14	24	0	0	92	2197	0.61
15	19	0	0	116	2207	0.61
16	17	0	0	141	2396	0.67
17	19	0	0	124	2356	0.65
18	17	0	0	136	2318	0.64
19	22	0	0	105	2319	0.64

voice calls blocked:	50	next voice channel:	14
total planned data hops	176	hop pointer on channel:	11
# times all channels busy:	15		
# individual busy periods:	15		
cumulative busy seconds:	111		

Blocked Second Histogram:

block scnds	numbr times	block scnds	numbr times	block scnds	numbr times	block scnds	numbr times
14	1	13	2	11	1	10	2
9	1	6	1	5	2	4	2
3	1	2	2				

	ENTERED	SIMULATED	>2% BLOCK
voice calls/hour:	369	377	
mean call length, seconds:	120	122 (average)	YES
arrival rate multiplier:	1	1	
peak period length (sec):	0	0	
number of peak periods:	0	0	

Priming Statistics:
elapsed time at start of 19th call: 164 seconds.
calls in process: 11

One Hour Run-Time Statistics:

channl	#calls ended	active calls	second to-go	second /call	total second	utili-zation
1	20	1	22	95	1902	0.53
2	19	1	66	142	2691	0.75
3	19	0	0	110	2094	0.58
4	19	1	51	133	2522	0.70
5	19	0	0	124	2357	0.65
6	20	1	139	133	2652	0.74
7	20	1	29	110	2208	0.61
8	21	1	97	118	2476	0.69
9	22	1	206	113	2490	0.69
10	21	1	3	121	2551	0.71
11	20	1	14	119	2382	0.66
12	19	1	50	127	2405	0.67
13	21	0	0	120	2523	0.70
14	18	0	0	130	2344	0.65
15	17	1	116	146	2480	0.69
16	19	1	107	132	2514	0.70
17	20	0	0	123	2452	0.68
18	20	0	0	124	2478	0.69
19	23	1	55	98	2265	0.63

voice calls blocked:	3	next voice channel:	13
total planned data hops	176	hop pointer on channel:	5
# times all channels busy:	7		
# individual busy periods:	7		
cumulative busy seconds:	27		

Blocked Second Histogram:

block scnds	numbr times	block scnds	numbr times	block scnds	numbr times	block scnds	numbr times
11	1	6	1	3	1	2	3
1	1						

```
           1         2         3         4         5         6  block
  123456789012345678901234567890123456789012345678901234567890   time

 1  000000000000000000000000000000000000000000000000000000000000    0
 2  000000000000000000000000000000000000000000000000000000000000    0
 3  000000000000000000000000000000000000000000000000000000000000    0
 4  000000000000000000000000000000000000000000000000000000000000    0
 5  000000000000000000000000000000000000000000000000000000000000    0
 6  000000000000000000000000000000000000000000000000000000000000    0
 7  000000000000000000000000000000000000000000000000000000000000    0
 8  000000000000000000000000000000000000000000000000000000000000    0
 9  000000000000000000000000000000000000000000000000000000000000    0
10  000000000000000000000000000000000000000000000000000000000000    0
11  000000000000000000000000000000000000000000000000000000000000    0
12  000000000000000000000000000000000000000000000000000000000000    0
13  000000000000000000000000000000000000000000000000000000000000    0
14  000000000000000000000000000000000000000000000000000000000000    0
15  000000000000000000000000000000000000000000000000000000000000    0
16  000000000000000000000000000000000000000000000000000000000000    0
17  000000000000000000000000000000000000000000000000000000000000    0
18  000000000000000000000000000000000000000000000000000000000000    0
19  000000000000000000000000000000000000000000000000000000000000    0
20  000000000000000000000000000000000000000000000000000000000000    0
21  000000000000000000000000000000000000000000000000000000000000    0
22  000000000000000000000000000000000000000000000000000000000000    0
23  000000000000000000000000000000000000000000000000000000000000    0
24  000000000000000000000000000000000000000000000000000000000000    0
25  000000000000000000000000000000000000000000000000000000000000    0
26  000000000000000000000000000000000000000000000000000000000000    0
27  000000000000000000000000000000000000000000000000000000000000    0
28  000000000000000000000000000000000000000000000000000000000000    0
29  000000000000000000000000000000000000000000000000000000000000    0
30  000000000000000000000000000000000000000000000000000000000000    0
31  000000000000000000000000000000000000000000000000000000000000    0
32  000000000000000000000000000000000000000000000000000000000000    0
33  000000000000000000000000000000000000000000000000000000000000    0
34  000000000000000000000000000000000000000000000000000000000000    0
35  000000000000000000000000000000000000000000000000000000000000    0
36  000000000000000000000000000000000000000000000000000000000000    0
37  000000000000000000000000000000000000000000000000000000000000    0
38  000000000000000000000000000000000000000000000000000000000000    0
39  000000000000000000000000000000000000000000000000000000000000    0
40  000000000000000000000000000000000000000000000000000000000000    0
41  000000000000000000000000000000000000000000000000000000000000    0
42  000000000000000000000000000000000000000000000000000000000000    0
43  000000000000000000000000000000000000000000000000000000000000    0
44  000000000000000000000000000000000000000000000000000000000000    0
45  000000000000000000000000000000010011111111111111111111111110   26
46  111111100011110111111110111101111111000000000000000000000000   30
47  000000000000000000000000000000000000000000000000000000000000    0
48  000000000000000000000000000000000000000000000011100000000000    3
49  000000000000000000000000000000000000000000000000000000000000    0
50  000000000000000000000000000000000000000000000000000000000000    0
51  000000000000000000000000000000000000000000000000000000000000    0
52  000000000000000000000000000000000000000000000000000000000000    0
53  000000000000000000000000000000000000000000000000000000000000    0
54  000000000000000000000000000000000000000000000000000000000000    0
55  000000000000000000000000000000000000000000000000000000000000    0
56  000000000000000000000000000000000000000000000000000000000000    0
57  000000000000000000000000000000000000000000000000000000000000    0
58  000000000000000000000000000000000000000000000000000000000000    0
59  000000000000000000000000000000000000000000000000000000000000    0
```

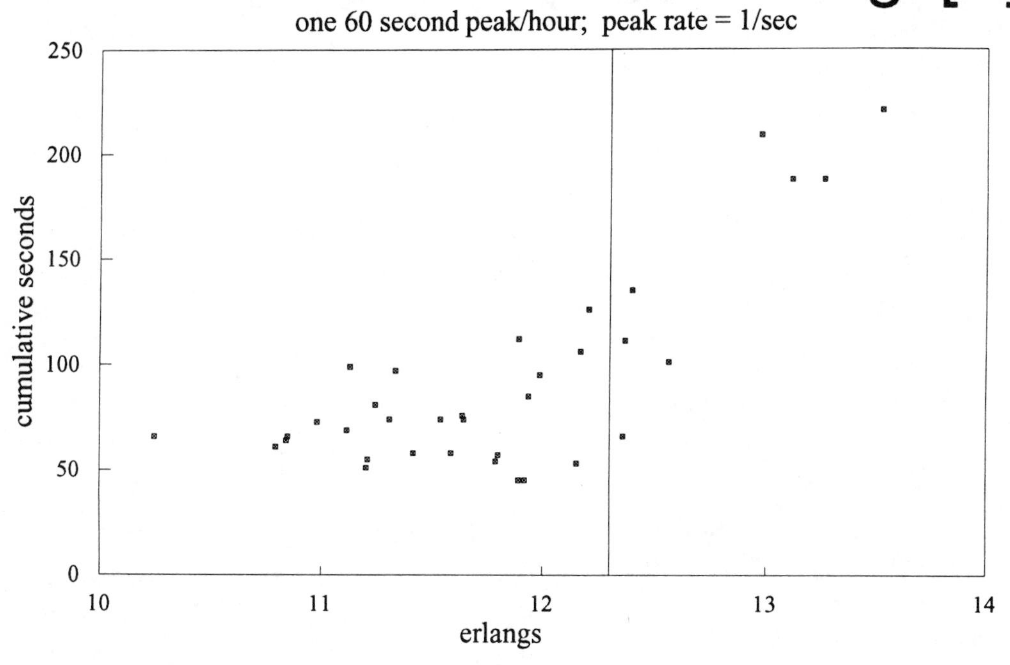

Comments:

1 - mean message length entered: 120 seconds
2 - Erlangs[B] = 12.3 for 2% blocking

INDEX

K

L

M

N